FEEDLOT EMPIRE: Beef Cattle Feeding in Illinois and Iowa, 1840-1900

FEEDLOT EMPIRE:

Beef Cattle Feeding
in Illinois and Iowa, 1840-1900

JAMES W. WHITAKER

A Replica Edition

The Iowa State University Press
Ames, Iowa
1975

For Mother and Father
    who have helped their sons beyond measure,
and the rest of the clan
    who have been interested and interesting observers

JAMES W. WHITAKER is Associate Professor of History, Iowa
State University.

© 1975 The Iowa State University Press
Ames, Iowa 50010. All rights reserved
Composed and printed by
The Iowa State University Press
First edition, 1975
International Standard Book Number: 0-8138-0625-9
Library of Congress Catalog Card Number: 74-14276

The Iowa State University Press REPLICA EDITIONS,
reproduced from typescript, are specialized studies selected
for their significance and scholarly appeal.

Feedlot Empire: Beef Cattle Feeding in Illinois and Iowa,
1840-1900, is based on a thesis submitted for the doctor of
philosophy degree in history at the University of Wisconsin,
Madison.

COVER: A turn of the century photograph illustrating the open
        lot method of cattle feeding. (Courtesy: State Histor-
        ical Society of Iowa)

CONTENTS

P R E F A C E

This book is concerned with the development of beef cattle feeding as a major agricultural industry during the latter half of the nineteenth century. The study is limited to
Illinois and Iowa because the industry reached its height
within these two political divisions. This is not to say that
man's political boundaries limited agricultural activity but
rather that, because agricultural data in the census are
gathered by and expressed in terms of political units, the
historian must take counties and states into account. Though
the industry described here went beyond Illinois and Iowa
boundaries into southern Wisconsin, western Indiana, and
northern Missouri, these areas are not discussed because it
is believed they would not significantly alter the patterns
that were developed.

The chronological limits of 1840 and 1900 are at logical
dividing points in the development of the industry. By 1840
men were transferring the pattern of cattle feeding in Ohio,
Indiana, and Kentucky to the prairies of Illinois and Iowa.
This movement was not new. From earliest Colonial times, men
had raised cattle on the fringes of settlement and driven
them to town and city markets. Carolina cowpens, backcountry
Pennsylvania, and western Massachusetts had provided cattle
for eastern markets; and as settlement moved westward, cattlemen moved further west. Soon after the American Revolution, cattlemen crossed the mountains into the Ohio River
valley to raise and drive cattle to Baltimore, Philadelphia,
New York, and even Boston markets. Pennsylvania farmers also
fattened Ohio cattle on corn before sending them to market.[1]

Once settlements crossed the mountains, the spread of
population forced cattlemen to move westward again to open
new grazing land. Ohio drovers appeared in Illinois in the
1820s and 1830s, and by 1840 cattlemen were well established
on Illinois prairies, raising cattle to be driven east. How-

vii

ever, when settlers reached the corn country of Illinois and
Iowa, instead of moving on in a few years they remained to
raise and feed cattle through the end of the century and to
the present day. Though men continued to carry the raising
of cattle westward to the plains, the center of prime fat-
cattle production remained in Illinois and Iowa.

By the end of the 1850s, cattlemen of central Illinois
and eastern Iowa had shifted from raising cattle to be driven
east for fattening in Ohio and Pennsylvania to fattening
their own on corn and grass for the Chicago and New York
markets. At this early date, the commercial outlines of Corn
Belt agriculture were visible. This important combination of
corn and livestock was present from the early nineteenth century. How to market the combination was the problem for the
farmers. Although the Corn Belt is often thought of as an
area of corn and hogs, it was and is just as much an area of
corn and cattle.

Cattle feeding in Illinois and Iowa started as a commercial enterprise; and contrary to the experience in earlier
areas, the business in these two states survived the technological changes and geographical shifts in areas of supply
and markets that took place during the nineteenth century.
By 1900, compared to the 1840s and 1850s, the industry had
changed in types of feeding routines, in quality of beef
produced, in age and weight of cattle at market time, and in
principal markets; it had evolved a new approach that remained well into the twentieth century. Through all this
change, Illinois and Iowa cattle feeders quickly moved from
the fringes of the fat-cattle trade in the 1840s to be the
leading prime-beef producers in the last half of the century.

This book is the story of the early development and the
subsequent growth of this distinctive aspect of Corn Belt
agriculture. External forces such as the development of the
railroad network, the rise of the urban market, and the development of refrigeration by the Chicago meat packers,
which completely changed the processing and marketing system
for fresh meat, had made changes possible and necessary in
the way cattle feeders operated. Illinois and Iowa men who
responded to the outside market influences by changing their
feeding systems to produce better quality beef at a younger
age led the way in the adjustment to new circumstances, which
kept these states the center of prime-cattle production.

This study could not have been completed without the

patient encouragement and help of Vernon Carstensen. Grateful
acknowledgment is also made to Morton Rothstein; to Allan
Bogue for sharing some statistics; to Iowa State University
Sciences and Humanities Research Institute, Graduate College,
and Department of History for financial aid; to the library
staffs of the State Historical Society of Wisconsin, the
University of Illinois, and Iowa State University; to the Illinois Historical Survey, especially its director, Robert
Sutton; and to Jan Miller for double-checking notes.

FEEDLOT EMPIRE: Beef Cattle Feeding in Illinois and Iowa, 1840-1900

C H A P T E R 1. ILLINOIS AND IOWA: PHYSICAL SETTING AND SETTLEMENT

The major portions of Illinois and Iowa, the core of that great agricultural area of the United States known as the Corn Belt, possessed an excellent combination of natural resources in soil, climate, topography, potential transportation routes, and lack of serious barriers to agriculture. Farmers quickly exploited this rich land and produced millions of bushels of corn as their major crop. Corn—a bulky, low-value grain—cost too much to transport far to market; but when fed to beef cattle and swine, it became a very profitable crop. This combination of prairie grass, corn, cattle, and swine early became the commercial base of Corn Belt agriculture; and though changes occurred during the nineteenth century, Illinois and Iowa farmers remained leading producers of fat cattle.

Corn grew on a wide variety of soils, but highest yields came from dark, well-drained, silt loam with an abundance of humus or organic matter. Humus was an important ingredient because it held moisture and prevented rain from running off or leaching the mineral elements too rapidly. Humus directly supplied nitrogen, the soil element needed in greatest amounts for corn.[1]

Geologic forces had endowed Illinois and Iowa with large areas of fertile soil ideal for producing corn and rich pasture. A series of glacial advances across the two states left them mantled with glacial till, crushed sedimentary rock (mainly limestone), wind-blown loess, and lacustrine deposits from glacial lake beds—all excellent soil-building materials. A good balance of acid and alkali in the soil south and west of Lake Michigan provided an excellent area for tall prairie grass. The deep roots and growth processes of this grass incorporated much organic material (humus) into the soil. This stored plant food, held moisture, fostered biotic life, and gave structure to the black prairie soils. Favored with a humid continental climate and warm summers, this land could produce in abundance for years with proper farming practices.2

At least four different glaciations combined to help create the bountiful soil of Illinois and Iowa. As each glacier advanced, it scraped off the high points and filled in the low spots. When the Wisconsin glacier (the last) receded, wind and water erosion shaped and modified the morainic and drift cover that remained. These elements created a terrain in the area of Illinois that was a gently undulating plain in the north, west, and south; nearly flat in the east-central part; and quite hilly and broken in the extreme northwest and in the south where the glaciers did not scrape off the tops of the Ozark hills. The Mississippi and Illinois rivers cut their channels below the level of the plain and in some areas created wide lowlands of rich alluvial or terrace soils. The area of Wisconsin glacial drift in the northern and eastern part of Illinois was the last of the state to be disturbed by glaciers and thus was exposed for a shorter period to the actions of wind and water. (See Map A.2. Map A.1 containing county outlines and names is the first of the maps in the Appendix.) Fewer mineral elements had been leached out of the soil by rain; therefore, the soil was more fertile than in other parts of the state. There also had been less time to develop a good drainage pattern for the area, with the result that successful cultivation awaited drainage projects. This was especially true of the flattest portion of the state in the east-central counties of Champaign, Ford, Iroquois, Kankakee, Livingston, and Vermilion.

During the last glacial recession wind-borne silt (loess), blown from the west, covered parts of the state. Thickest in the western part, it thinned out to the east and south. Variations in loess cover, age of the glacial drift, and degree of weathering accounted for important variations in Illinois soils and consequent agricultural productivity.

Most of western Illinois was covered with loess four feet or more thick. Exceptions were the alluvial area of the lower Illinois River valley; sandy, glacial drift and wind deposits in Mason County; and a basin in the north, covering parts of Whiteside, Lee, Bureau, and Henry counties. The loess and alluvium areas were very fertile, but the sandy areas were less so and subject to poor drainage. Wind depos-

4

ited very little loess upon the fertile drift cover left by the last glacier in the northeast quarter of Illinois. In the eastern counties of Kankakee, Iroquois, and Ford were some less fertile sandy deposits. Just south and west of this area loess existed from two to four feet thick in the east-central part of the state; the southern part south of Coles County had only small deposits. Distribution of the fertile loess duplicated many of the other fertility producing factors, and its absence from most of southern Illinois decreased that area's prospects for raising corn. This was one of the reasons that southern Illinois farmers did not concentrate on beef cattle production.[3]

The vegetation that developed on top of this soil material modified it. By the time of American settlement prairie grass covered much of Illinois north of Adams County west of the Illinois River and north of Christian County east of the river except for deciduous tree cover along the watercourses and in the extreme northwest. South of Adams and Christian counties tree cover was the rule and occasional prairies the exception. The dark prairie soils of northern and central Illinois, further enriched by generations of grasses, were more productive than the lighter forest soils of southern Illinois and the more broken country. Much of southern Illinois had a layer of hard, relatively impervious subsoil known as claypan, which prevented good drainage, reduced absorption of rainfall, restricted the rise of subsurface water, and favored development of shallow root systems. The presence of claypan became critical during wet and dry spells by either keeping the topsoil too wet or letting it dry out too quickly. (See Map A.3.) Forest soils (other than that of the claypan area), though less fertile than prairie soils, were nevertheless suitable for a wide variety of crops.[4]

Iowa topography was a gently rolling plain with the highest elevations in the northwest. The roughest, most broken terrain with the highest percentage of land not suited for cornfields was in the northeast corner and comprised Allamakee County and parts of five nearby counties. After the action of the same four glaciers that affected Illinois, Iowa had no area of the claypan soil that hindered southern Illinois agriculture. During the retreat of the Wisconsin glacier winds covered over two-thirds of present Iowa with four feet or more of rich soil-building loess. North-central Iowa from Dickinson County south to Dallas County and northeast to Franklin County received no loess cover during the last glaciation (as did east-central Illinois) and was the flattest portion of the state. Thus it had the most drainage problems, making it suitable first for grazing country rather than grain crops. With the exception of the counties adjoining the Missouri River in the west and the river valleys in the eastern part of the state, most of Iowa was prairie. Combined with loess and the favorable glacial drift, centuries of biotic action by the prairie grass created the brunizem prairie soils over most of Iowa, perhaps the "most productive soils known to man."[5]

Corn Belt climate was as favorable as the soil resource. All Illinois and Iowa had an average growing season of over 145 days from the last killing frost in spring to the first in fall. Varieties of dent corn dominant in Illinois and Iowa in the last half of the nineteenth century needed from 90 to 125 days of tolerable temperature and rainfall conditions to mature. Corn grew best with hot wet weather in July and August and in this, too, climate favored the Corn Belt. Most of the yearly precipitation for Illinois and Iowa came in the warm months with an average of 20 to 25 inches of rain between April 1 and October 31. On the average at least half of this rain came in the three crucial months of June, July, and August. Periodic droughts and variations in the average dates of frosts occurred, causing variations in the crop yield; but natural conditions usually favored agricultural productivity.[6]

In addition to soil fertility and climate, natural transportation routes for movement of people and produce favored agricultural development in Illinois and Iowa. The drainage systems created by glacial runoff and subsequent centuries of weathering formed the Great Lakes and the Ohio, Illinois, Mississippi, and Missouri rivers, which with their tributaries made it possible to till much of the land without artificial drainage and also provided a low-cost transportation outlet to markets to the south and east. The low elevation of the divide between Lake Michigan and the Illinois River permitted construction of a canal linking the two, and the relatively level terrain made it easy soon after to connect lake ports by railroad to the productive agricultural interior.

Physical factors made most of the rich land of Illinois and Iowa fully able to produce an abundance of food for men and animals if farmers put it to use. These factors also differentiated the area and in part accounted for different types of agricultural uses. Well-drained fertile prairie soil was good for grains, especially corn; poorly drained prairie served best in early years as pasture for cattle. Claypan soils inhibited corn production. Rough terrain was better suited to pasture crops than grain and encouraged dairying. Nevertheless, the past experience, current predilections, and ability of settlers on this rich soil accounted in an unmeasurable but important way for the use they made of available resources.[7]

Westering Americans (and immigrants) did not take long to learn of, covet, and possess Illinois and Iowa land. People came by ones and twos, by families and towns, by foot and horseback, by wagon and canal, by lake and riverboat. By the early 1830s the Erie Canal and the Ohio River had opened the gates to a flood of humanity that surged west from Lake Michigan and up the valleys of the Wabash, Big Muddy, Kaskaskia, Illinois, Sangamon, and Rock rivers in Illinois and started up the Des Moines, Skunk, Iowa, Cedar, Wapsipinicon, Maquoketa, and Turkey river valleys in Iowa.

Enticing reports of the fertile and beautiful country had been appearing for years. Colonel George Rogers Clark, reporting on his military campaigns during the Revolution, in 1779 described southern Illinois as "more beautiful than any idea I could have formed of a country almost in a state of nature." He predicted that someday "meadows, extending beyond the reach of your eye," would "excel in cattle."[8]

Scottish farmer, Patrick Shirreff, after his travels through Illinois in 1833, pronounced it "not greatly surpassed by any portion of the earth" for agricultural quality and potential and predicted that it would have "sufficient room for all who are likely to desire a settlement." Two years later Albert Miller Lea writing of his reconnaissance with the United States Dragoons heralded yet another "Eden" in the Iowa River valley, which "presents to the imagination the finest picture on earth of a country prepared by Providence for the habitation of man." Many other travelers and residents testified to the agricultural potential of the region. These early estimates of the productivity of the area received a back-handed seconding from Horace Bushnell, Connecticut minister, lecturer, and essayist, who in 1846 labeled the incipient Corn Belt, not as a beautiful, hilly paradise such as New England, but rather the visually unappealing "great American Corn-field, the Poland of the United States."[9]

By 1846 Iowa had garnered enough people to become a

state even though Indians still held part of it; and Illinois, though a state since 1818, showed a remarkable ability to absorb all those who desired to settle. By 1850 only the Grand Prairie area of Champaign, Ford, Iroquois, Kankakee, Livingston, and Piatt counties and the wet bottomland of Bureau and Lee counties had less than six people per square mile.[10]

The transition from Indian country to statehood went quickly once settlement had begun. Beginning with the first Indian land cessions (1795 in Illinois and 1824 in Iowa), an evolution of events followed: federal land survey and sales, increasing settlement, and the development of political structures. Indians ceded land in Illinois from 1795 to 1833 and in Iowa from 1824 to 1851. Federal land surveyors had finished their work in Illinois, for all practical purposes, by 1844 and in Iowa by 1861.[11]

Early settlers urged their friends and relatives to come to "the flower of the World, the Garden of Eden, a land flowing with milk and honey." From these passionate urgings, from general reports about soil fertility, and from an assumed salubrity of climate as well as the many common motivations for human migration, settlers flocked to Illinois after 1815 and to Iowa after 1833. They moved up the river valleys first where there was a transportation route and timber for fuel and shelter and then pushed on to the prairies, and rapid population growth continued. By 1900, 25.6% of Iowans and 53.9% of Illinoians were classified as urban dwellers in towns of 2,500 population or more. (See Table B.1.) This meant that an increasingly smaller percentage of the population accounted for the growing agricultural production. Illinois had 264,151 farms in 1900 and Iowa had 228,622, but from these farms came an abundance making the two states leaders in farm production.[12] (See Table B.2.)

As people moved into this region, they quickly shifted farming from a pioneering way of life to a business. Corn and livestock thrived on the rich land and favorable climate from the beginning of settlement; and although distractions appeared from time to time, corn and cattle remained major products. By the end of the century, they accounted for the largest percentage of the value of farm products of the two states.[13] (See Table B.2.)

CHAPTER 2. ILLINOIS-IOWA
AGRICULTURE IN THE NINETEENTH CENTURY

As farmers settled in Illinois and Iowa in the 1840s and 1850s, they sought to develop a commercial agriculture. Rather than attempting to be self-sufficient by producing a wide variety of crops and livestock, frontier farmers quickly concentrated on a few income-producing activities, which they changed as market conditions changed. No exception to American agricultural experience, these men made little attempt to conform to a literary ideal of the self-sufficient pioneer. They simply were not interested; the long hours, versatile talents required, and lack of opportunity for the capital accumulation necessary for an attempt at self-sufficiency did not appeal. The productivity of the land, the options before them, and the developing transportation system encouraged farmers to think in terms of commercial rather than subsistence agriculture.[1]

Men turned from the sod-corn and wheat of the pioneer years to cash-grain and livestock production. This was a quick transition in central and northern Illinois and eastern Iowa with less experimentation with a variety of crops than took place in Wisconsin, for example. There farmers spent two decades growing wheat, sugar beets, sorghum, hops, flax, tobacco, and sheep before they found dairying to be the best combination of resources for commercial agriculture.[2] No need existed for this long period of searching for commercial combinations in Illinois and Iowa where the grass and abundance of corn made livestock feeding a logical choice.

Distractions at times drew some farmers away from cattle feeding, but these did not last. In the 1840s high prices for wool encouraged many Illinois farmers to raise sheep, with a resulting increase in wool production from 1,500 pounds in 1842, to 246,610 pounds in 1846, to an estimated 1,200,000 pounds in 1849. Sheep raising declined in popular-

9

ity during the 1850s but increased again in Illinois during the Civil War because of the cotton shortage. After the war, however, wool prices dropped and farmers turned to more rewarding enterprises. Iowa, too, experienced the sheep boom and the disillusionment in the late 1860s. In Cedar County, for example, "the sheep mania . . . proved a bad speculation." As an alternative to sheep or the unrewarding wheat, men called for more "capital and labor engaged in dairy, and stock raising."3

Chinese sugarcane (sorghum) and sugar beets were two other minor enterprises tried in the 1850s and 1860s. By 1860 Iowa accounted for 18% and Illinois for 12% of gallons of sorghum produced in the United States.4 The first successful Illinois sugar beet refining operation began at Chatsworth in Livingston County in 1867 and processed 40 tons a day in 1868. Two years later a Shelby County farmers' club issued an "Invitation to Capitalists" to establish a sugar beet factory there. The Chatsworth factory moved to Freeport in 1871 to be closer to land better suited for growing beets.5 But neither sorghum nor sugar beets was ever a major crop in acreage or value of product.

A surprisingly short time elapsed after settlement until Illinois and Iowa farmers turned to commercial production of grain and livestock. Little effort was required to plant corn and even less to harvest it because cattle could be turned into the fields and allowed to harvest the crop themselves. What the cattle knocked down and wasted, swine could salvage. Corn was the basic survival crop for both men and animals, so that corn and cattle were present from the time of settlement and only gave settlers the problem of finding their most profitable use. Henry R. Schoolcraft, United States Indian Agent and member of several reconnaissance expeditions in the upper Mississippi valley, remarked on this production and the potential of an outside market in Sangamon County, Illinois, as early as 1821. Schoolcraft thought the region was prepared to compete with Michigan and Ohio in provisioning government posts on the Great Lakes. Timothy Flint, missionary to the Mississippi valley and writer, reported in 1828 that eventually prairies would be habitable after people planted trees on them, but in the interim the "inexhaustible" summer range for cattle gave Illinois an advantage over eastern and southern states in the cattle trade. Bluegrass, said Flint, had appeared in areas where settlers had killed prairie grass, and the state "sent great numbers of fine cattle and horses to New Orleans." Five years later in 1833 Illinois farmers had plowed under sufficient prairie for grain production to warrant Patrick Shirreff's observation that forest agriculture "is seldom followed, the cultivation of the prairie being so much more simple and profitable." Farming seemed easy to Shirreff in terms of labor and capital investment, and some settlers attested to this in noting that sickness was no "great misfortune" because it was so easy to make a living. An Iowan reported the land so rich he could raise as much with half the labor as in Ohio. Others disagreed, at least to the extent of saying that it was a hard country for women and cattle. Two universal comments, complaints that applied well into the latter half of the century, were that farmers did not provide adequate shelter for livestock and farms were too large to be worked efficiently.[6]

From 1830 to 1860 wheat was a major money crop, but after the 1830s, increasing amounts of corn, cattle, and swine were produced and exported from the area in some form. With the exception of southern Illinois, the two-state area justified the name Corn Belt before the Civil War. In 1859 a Knox County, Illinois, farmer exclaimed: "Corn, the country is one immense corn field."[7]

By the time of the 1850 census one-third of the land in Illinois, with less than 900,000 inhabitants, was already in farms. (See Map A.4.) But the land was so productive that Illinois farmers (with only 5.2% of the farms and 4.4% of the improved farmland in the nation) produced an abundance of corn and wheat disproportionate to their numbers and a slightly greater proportion of oats, "other" or beef cattle (i.e., not dairy cows or oxen), and swine (which followed the cattle in the feedlot during corn feeding at the approximate rate of two swine for each steer). Iowa lagged about 10 to 15 years behind Illinois in the growth of agricultural production, and in 1850 only 7.7% of the land in the state was in farms. While the less than 200,000 inhabitants of Iowa held 1% of the nation's farms and only 0.73% of the nation's improved farmland, production of wheat and corn exceeded a proportional share of the national total.[8] (See Table B.3.) Corn and livestock were already a major part of Corn Belt agriculture.

Production totals for the two states do not tell the whole story, for in 1850 concentrations of the various crops existed within each. Illinois farmers in northern counties

11

close to lake ports or central counties along the Illinois River and Illinois-Michigan Canal could raise wheat as a cash crop after 1848. East-central and southern Illinois farmers concentrated more on corn for the southern market or on local livestock. Swine were distributed rather evenly over the state, but there was a concentration in the corn counties (except De Witt and Vermilion-Ford); in the southeastern counties from Crawford to Williamson; and in the west-central counties of Knox, Warren, McDonough, and Cass. Counties north of Marshall had the least number of swine. "Other" cattle were most plentiful in the central counties of Cass, Greene, Macoupin, Menard, Morgan, Sangamon, and Scott with lesser concentrations in a belt from Edgar through Vermilion-Ford, McLean, Tazewell, Fulton, McDonough, and Adams and in Madison and Randolph counties east of St. Louis. The corn-cattle-hog pattern of Corn Belt agriculture already existed though it was not yet highly developed.[9]

The decade 1850 to 1860 saw many changes. After several years of good crops, some improvement in transportation, and good prices because of the Crimean War, wheat became less reliable in northern Illinois. Crop failures because of spring rains, late frosts, and insects after 1855 caused many farmers to take to heart the calls from the agricultural societies and periodicals to give up wheat for more stock raising. By 1860 northern Illinois farmers placed considerably less emphasis on wheat, a condition that lasted through 1870. Farther south, wheat cost more to produce than it was worth in some inland areas; some even considered it a "nuisance" that just bred chinch bugs. More often than not, standing water killed winter wheat in central and eastern Illinois in areas of poor drainage. The drill, which provided a better method of planting wheat; a shift to winter wheat, which could be harvested earlier in the season; and, most significant, the coming of railroads to southern Illinois in the 1850s caused an increase in wheat production in the south. In 1870 Commissioner of Agriculture Horace Capron felt it necessary to devote a major paragraph in his speech at the Illinois State Fair to urge farmers to depend less on wheat. The total wheat acreage fluctuated between 1.7 and 2.5 million acres from 1859 to 1874; but as total acres in farms increased, wheat became proportionally less important.[10]

Because of soil and terrain and, to some degree, because of the type of farmers who stayed there, southern Il-

linois did not share in the development of the corn-livestock
farm economy of the rest of the area. Though the state was
rather backward in Corn Belt farming techniques and attitudes (a reaper salesman called southern Illinois farmers
the "coon dog and butcher knife tribe"), an effort to excel
in horticulture was successful. As early as 1836 Gershom
Flagg had 500 apple trees on his Madison County farm, and a
major part of his income came from fruit sales; but interest
in large-scale production did not come until the 1850s and
later. When production increased, the Illinois Central Railroad operated an express train to carry peaches, apples, and
berries into Chicago. By 1868 one station in Shelby County
claimed it had shipped 50,000 boxes of peaches in one
season.[11]

From 1860 on, in response to the growing urban market,
northeastern Illinois included more dairying and truck garden production. Gail Borden opened his first milk condensing
factory at Elgin in 1865; and by 1870 Kane, McHenry, De Kalb,
Cook, Lake, Du Page, Kankakee, Boone, and Winnebago counties
were important cheese-producing centers.[12]

The rise of dairying was only one of the signs of growing regional specialization in farming. In the 1850s east-
central Illinois turned to corn and livestock when the Illinois Central Railroad opened the area by bringing in settlers
and providing transportation to market. Farmers in counties
close to the Mississippi, Illinois, and Ohio rivers had
shipped corn, beef, and pork to the south; but the Civil War
blockades interrupted this, forcing men to seek markets in
the east by way of Chicago. The higher cost of freight (in
proportion to the value of grain) by wagon and railroad to
Chicago caused many farmers to concentrate on feeding corn
to cattle and swine. Livestock could walk to the nearest
railroad, and freight charges on livestock were a lower percentage of the value of the product than charges on grain
shipments.[13]

In the meantime Iowa farming went through changes similar to those in central and northern Illinois. Though by
1860 not nearly as much of Iowa as Illinois had over 50% of
the land divided into counties in use as farms, the grain-
livestock economy was well developed in the eastern part of
the state. (See Map A.4.)

Early settlers had planted wheat and corn in the Cedar,
Iowa, Des Moines, and other river valleys. No area corresponded to southern Illinois and its problems. The river sys-

tem channeled trade to the south until the railroads extending west from Chicago crossed the Mississippi. By 1860 four Iowa railroad lines met Illinois railroads from Chicago at the river. In 1850, for example, Wapello County, about 70 miles up the Des Moines River, had over 51% of its land in farms and over 35% of the land was improved. Of the 978 householders for the 1850 census, 178 came from Illinois, 369 from Indiana, and 242 from Ohio, so that most were familiar with farming as practiced in the old Northwest.14

Some areas of agricultural concentration existed in Iowa in the 1850s. Eastern Iowa counties north of the Cedar River and Benton, Iowa, Johnson, Mahaska, Louisa, Des Moines, and Lee counties concentrated on wheat. (See Map A.1.) Muscatine, Taylor, and Page counties and the area from Keokuk to Madison county and south embraced the higher corn-producing areas. The swine concentrations centered in the corn counties and in Marshall, Fremont, Washington, and Des Moines counties. Cattle concentrations occurred in widely scattered counties such as Allamakee and Delaware in the northeast; in Pottawattamie, Fremont, Page, Taylor, and Decatur in the southwest; and in Clinton, Muscatine, Louisa, Des Moines, Lee, and Washington in the southeast. Many Iowa farmers also had turned to corn, cattle, and hogs.15

But as the corn-cattle pattern appeared, changes in agricultural technology encouraged men to continue to return to grain production. The periodic lure of corn or wheat as a cash crop continued to attract farmers away from cattle feeding during the 1850s, 1860s, and 1870s. Through a combination of availability of railroads, type of land tenure, cost of drainage, and price of beef in the 1880s and early 1890s, east-central Illinois was fully and finally turned toward producing cash grains rather than fat cattle. Most important in encouraging grain production were the increased invention and use of agricultural machinery in both states from 1840. Perhaps most dramatic in their impact on agriculture were the plow and the reaper. Prairie plows opened up more new land, and the reaper made larger wheat and oat fields per farm possible. By 1851 the McCormick factory produced over 1,000 reapers a year at Chicago, and John Deere turned out 10,000 plows a year by 1857. In 1859 one man in Piatt County, Illinois, had a portable steam engine, four threshing machines for steam or horse power, a corn sheller, and more work for them than he could handle. During the last half of the century, mechanical innovations continued as inventors refined tillage machinery, such as Oliver's introduction of the chilled-iron plow in 1870, and harvesting equipment to make each man-hour more productive.[16]

Of great importance in aiding the expansion of farms across the two states was the growth of the transportation network. Railroads played the most important part in marketing the agricultural produce of the interior counties and providing an all-weather outlet to the east. From only 788 miles of track in Illinois and none in Iowa in 1854 to 4,823 miles in Illinois and 2,683 miles in Iowa in 1870, the railroad networks spread out from Chicago. By 1900, Illinois had 11,002 miles and Iowa had 9,185 miles of single-track line.[17]

Innovations in machinery, development of the transportation network, and growth of population caused the number of farms in Iowa and Illinois to increase from 1850 to 1860 by 313% and 88% respectively and from 1860 to 1870 by 90% and 41%, while the United States increase in the two decades had been only 41% and 30%. (See Map A.4.) The amount of improved land in Illinois increased by 159% from 1850 to 1860 and by 47% from 1860 to 1870. In Iowa improved land increased by 359% and 147% for the same periods. Thus by 1870 farmers had turned 72% of the land in Illinois into farms and 43.7% of the land in Iowa. Production figures for 1870 compared to 1850 indicate the same quick growth in a rich area and reveal the dominance of feed grains over wheat in Illinois. (See Table B.4.) The feed-grain dominance was less pronounced in Iowa than in Illinois in 1870 than it would be later, but the trend was established.[18]

From 1870 to 1900 further regional shifts occurred in agricultural production within Illinois and Iowa, but the result was more corn and less wheat. As settlement moved westward in Iowa and the transportation network developed, Iowa agricultural regions changed. The center of corn concentration moved toward the southwest, the size of the area where sheep were important became smaller, and farmers in the northeast counties moved toward more dairy production. East-central Iowa farmers shifted toward meat production of swine and beef fed on corn, and those in the north-central counties began raising more oats in addition to the dairy and beef cattle. Wheat production centered in the northeast. Iowa was moving toward the early twentieth-century divisions of northeastern dairy area; north-central cash-grain area; western corn, beef, and swine area; south-central and southeastern pasture areas; and east-central livestock area, with swine predominating.[19]

After 1870 Illinois farmers in the northeastern counties

15

around Chicago concentrated increasingly on dairy and truck produce. Those in the northwest counties, having found a good combination, continued the trend toward cattle and hogs. Farmers north and west of the Illinois River continued in livestock and grain, but men in the east-central region turned increasingly to cash-grain operations in place of cattle feeding. Farmers in the Sangamon River counties, who had raised both wheat and corn, turned more to corn. From Shelby County south, operations could be called "general farming" where there was a combination of corn, wheat, hay, cattle, and swine. Around East St. Louis some concentration of dairy cattle and produce existed and there were areas of fruit specialization in the southern counties; but on the whole the southern fifth of the state was out of the main trend of Corn Belt agriculture.[20]

By 1900 the two-state area, which had only 3.7% of the land surface of the United States, contained 13.9% of the improved farmland of the nation. Most farmers had given up raising wheat and turned to feed grains and livestock, which had been indicated as the major commercial combination as early as 1850. (See Table B.5.) Corn was the most important income producer of the grains, accounting for 54.2% of the value of all farm crops produced in 1899 in Illinois and 50.6% in Iowa. Oats accounted for just over 17% of cereal crop values in each state, and wheat contributed less than 6%. Of the value of domestic animals on farms in June 1900, "other" or beef cattle (i.e., not dairy cows 2 years old or over) accounted for 25.6% in Illinois and 35.3% in Iowa. Swine, though in larger numbers on farms, accounted for only 12.6% of the value of domestic animals in Illinois and 16.0% of that value in Iowa. Sheep in each state accounted for less than 2% of domestic animal values. The cattle and swine figures do not take into account the number of animals marketed annually, so they are not completely representative of the two-state share of cattle raised and marketed during the year. Because many of the hogs marketed came as an adjunct to cattle feeding, one can discuss cattle as the major livestock interest of farmers feeding corn without excluding swine. Surely the man prophesied rightly who had described this area as "the great bank from which our sustenance is discounted."[21]

This concentration of the Corn Belt on feed grains and livestock was taking place at a time when the population of the United States continued to grow at a fast rate. More significantly, an increasingly greater proportion of the population was urban and did not produce its own food. The growing urban areas provided Corn Belt farmers with the opportunity as well as the need to change their agriculture from simply a way of life to a highly organized commercial operation. By 1900, 37% of the United States population lived in places of 4,000 or more residents, and the trend was toward greater urbanization. The 1900 census figures represented a 20% increase over 1890 in total population, and a 36% increase in urban population over 1890. For the same period the population of Illinois increased by 26%, but its urban population increased by 53% for a total of 51% of the state living in places of 4,000 or more. Iowa, much less urban, gained 16% in total population and 36% in urban dwellers from 1890 to 1900, when 20% of the state lived in places of 4,000 or more.[22]

The land of the Corn Belt was suited to produce for the growing urban market by geography and the type of farmers who settled it. Nowhere was this response to that market more noticeable than in the changes that took place in the cattle feeding industry in Illinois and Iowa. Feeding corn to cattle was well established by 1850 in the Corn Belt; but by 1870, in response to the expanding urban market, men devised refrigeration methods to ship fresh beef to any part of the country. In turn the demand for better quality meat for the refrigerated dressed beef trade brought changes in methods of feeding and in the quality of beef cattle in Illinois and Iowa.

C H A P T E R 3. ILLINOIS-IOWA BEEF PRODUCTION TO 1860

Corn Belt farming is perhaps most often thought of as a concentration on corn and hogs, but equally important in the nineteenth century was the combination of corn and cattle. Feeding corn to cattle and then using hogs to salvage the waste was a successful commercial agricultural pattern in central and northern Illinois and eastern Iowa.

In the years before the changes of the 1860s and 1870s, Illinois and Iowa cattlemen developed a pattern of operations similar to that of areas in Ohio, Indiana, and Kentucky. From Colonial times, cattlemen operated on the fringes of the settled area where land was cheaper and animals could graze over large areas. A range cattle business developed in western Massachusetts, in central Pennsylvania, and in the backcountry of the South. Men collected their cattle periodically and drove them to city or town markets. As the area of settlement expanded in the late eighteenth century, Pennsylvania and Virginia cattlemen moved their operations over the mountains into the Ohio River valley. There they found the combination of corn and grass in central Ohio and Kentucky ideal for raising and feeding cattle. Livestock feeding was also a way to market the frontier corn crop, which was not profitable to ship any distance as grain because of high transportation costs. However, when distilled into whiskey or fed to cattle and hogs, corn became profitable. Ohio and Kentucky farmers drove corn-fattened cattle over the mountains to market. As early as 1805 George Renick had driven cattle from southern Ohio to Baltimore, and by 1820 cattle driven over the Wilderness Road, Cumberland Road, and Mohawk-Erie Trail supplied the eastern markets. The rich farming counties in eastern Pennsylvania specialized in finishing Ohio cattle for the Philadelphia and New York butchers. In the middle Scioto River valley in Ohio, bluegrass country in Kentucky, and upper Wabash River valley in Indiana, farmers increasingly specialized, first on raising and driving cattle east, then on fattening those obtained farther west.[1]

By 1819 Ohio cattle feeders had traveled as far west as Missouri and Illinois in search of young animals to return to Indiana, Ohio, and Kentucky for fattening on corn in preparation for the overland drive to eastern markets. Some of the cattlemen soon took advantage of the rich prairies of Illinois and Iowa as places to raise cattle and corn more cheaply than in Ohio or Kentucky. By the 1820s and 1830s, even before the complete removal of the Indians, travelers commented upon the lush prairie grass that made Illinois a natural grazing area. They also noted that Illinois men already drove cattle to eastern markets or sent them to New Orleans. By 1831 in Madison County "cattle raising was carried on at a large scale"; in Morgan County settlers remembered that in the "early years," the great profits came from raising and feeding cattle; and by 1833 in the area around Springfield, Illinois, there were "a great many cattle . . . reared on the prairies."[2]

In addition to supplying stock to Ohio corn raisers, Illinois man fattened cattle on prairie grass, bluegrass pasture, and corn. In the early 1830s, Patrick Shirreff reported that farmers in central Illinois complained that cattle did not "thrive well" on cured prairie grass. He reported that the more experienced farmers used prairie grass only for summer grazing and planted timothy for cutting because it yielded a ton more per acre than either bluegrass or red clover. Farther north in Putnam County, by 1849 farmers replaced prairie grass with timothy because the wild grass was not ready for grazing until late spring and failed earlier than timothy in the fall. Also, increased use of fencing prevented men from driving cattle to open prairie and back every day.[3]

Drovers brought the cattle feeding pattern to Illinois when they returned to settle after seeing the advantages of cheap prairie grass and corn there. They changed from being itinerants in the west to settled landowners and made the adjustment to a new kind of operation. Such men as John T. Alexander and Jacob Strawn of Morgan County, Isaac Funk of McLean County, and Benjamin Harris of Champaign County set the example for Corn Belt agriculture as they made the cattle feeding operation work in Illinois. In a few years they began to compete with Ohio cattlemen in driving fat cattle from Illinois to eastern markets. Benjamin F. Harris, for example, was born in Virginia in 1811 and moved with his parents to Ohio in 1833 where he worked as a drover taking stock from Illinois, Indiana, and Ohio to the eastern markets. He came to Illinois as early as 1835, and by 1841 he grazed large numbers of cattle on the prairies of east-central Illinois. When he found little market for the grain he also raised, he developed a major corn feeding operation.[4]

Harris was just one of the many who moved west and brought familiar cattle operations with them. Agricultural developments in Illinois and Iowa had begun to follow the pattern of transition from range country to a grain growing and cattle feeding region. By the 1850s eastern Iowa men fattened cattle for Chicago and eastern markets, and southwestern Iowa around Council Bluffs and areas in northern Iowa supplied feeder cattle for eastern Iowa and Illinois. Increased land values in Ohio, rising prices for corn, and lower costs of cattle production in the states to the west gradually aided the shift in the emphasis on feeding from Ohio to Illinois and Iowa.[5]

Agricultural writers such as Solon Robinson (Indiana farmer, frequent correspondent to the Prairie Farmer, and occasional agricultural editor of the New York Tribune) felt that not enough farmers took advantage of natural conditions in Illinois and Iowa favoring cattle feeding. In 1850 Robinson wrote: "Now, my friends, is it not time for you to begin to think that wheat is not the most natural and profitable staple crop of this part of Uncle Sam's big pasture? Does any land in the world produce better beef than the prairies of Indiana, Illinois, Wisconsin, and Iowa? Grass, either wild or cultivated, is ever growing luxuriantly upon an inexhaustible soil. Indian corn, the best crop in the world for beef, rarely, if ever fails." Not alone nor the first to say so, Robinson, one of the more eloquent and important agricultural writers of the day, saw the possibilities for the prairie states. His 1850 letter to the Prairie Farmer came after another wheat failure; he then argued that the Middle West should concentrate on producing cattle because they were more profitable than wheat. Maryland and Virginia, he argued, could raise wheat successfully but could not raise cattle as cheaply as the Middle West with its prairies for grass and corn. Robinson hoped the wheat failure would "induce the people of all this great grass growing region . . . to sit down and count the cost" of growing poor wheat crops against raising cattle.[6]

Charles Mason (United States Commissioner of Patents and an Iowa farmer and stock raiser) heartily seconded the idea of switching from wheat to livestock, claiming that "the great defect in our agriculture is the failure to rear the proper number and quality of animals." In addition, men argued that use of cattle manure would increase the total grain crop by continuing a high state of fertility on cropped fields. Cattle also could walk to market and could consume surplus grain in a profitable way, and prices for good animals were more stable than fluctuating grain prices. Comparing all this to wheat failures and the cost of transporting grain to markets, many saw cattle feeding as the logical enterprise.[7] It was only about twenty years since Ohio drovers had settled in Illinois, but by 1856 enough farmers marketed fat cattle to have Illinois beef frequently make up the largest percentage of reported western cattle offered in the New York markets, thus surpassing Ohio. This rising Illinois business even caused problems in Ohio where cattle fattening declined in some areas that were formerly dependent upon feeder cattle from Illinois.[8]

By the time of the Civil War, cattle feeding operations of a similar type were well established in central and northern Illinois and eastern and southern Iowa. Abundant prairie beyond settlements provided range and corn land. Range grazing of cattle moved west and south, though some remnants remained on the wet prairies in east-central Illinois and in north-central and western Iowa until the 1880s. Before 1860 men had difficulty finding enough feeder cattle. Pioneers kept a cow and a few work animals, often turned loose on the prairie, but few were willing or able to sell more than a calf or two or a dried-up cow to a drover. Men rode far and wide in search of feeder cattle.

Isaac Funk, after learning the business in Ohio, moved to central Illinois in 1826 in order to have sufficient rangeland, but he found his operation limited by the small number of cattle in McLean County. This forced him to travel into Sangamon County to find settlers willing to sell. As settlement increased, so did the number of available feeder cattle and the magnitude of Funk's business. Eli Strawn had a system of subsidiary buyers in counties around his Ottawa, Illinois, headquarters who bought feeder cattle and collected fat-cattle herds for him. Others went further afield than Funk and Strawn.[9] Large-scale operators gathered their herds from farmers who felt that feeding cattle was not profitable

21

unless they had two dozen head or so.10 These feeders usually did not raise their own calves, so of necessity they developed a discriminating eye for estimating the current weight and fattening potential of calves they inspected.

In search of larger supplies of relatively low-cost feeder cattle, cattlemen reached out to Cherokee country (Oklahoma), Texas, and even New Mexico. This movement of western range cattle into the Corn Belt for fattening was a repetition of the drover pattern that had been connected with cattle feeding in Ohio. Here men repeated the three stages of the cattle feeding business: a progression from range country to feedlot to market. The range had moved from the Colonial backcountry to Ohio, Illinois, and Iowa, and then onto the plains. In the early 1850s, Illinois and Iowa cattlemen became aware of the possibilities of feeding cattle obtained from the plains. Men from Piatt County, Illinois, fed Texas cattle in 1852; and in 1853, Tom C. Ponting drove a herd to Illinois and sold them in New York in 1854, supposedly the first such cattle entered in the New York market. The next year cattle from Cherokee country and Texas passed through Springfield on their way to eastern Illinois feeders. Also in 1855 two Illinois men drove a herd from Texas to Iowa; a Chicago newspaper reported that a man named McCoy, operating from La Salle, was driving a herd of Texas cattle to Illinois. The next year, an Illinois man reported a profit of $24 a head on Texas cattle sold in New York. Encouraged by these prospects, Corn Belt cattlemen shipped to the Allerton Stock Yards in New York City each year from 1856 to 1859 an estimated 1,000 to 2,000 head of Texas cattle fattened on grass or corn in Illinois and Iowa.11

After the Civil War, feeders returned again to Texas and the plains for cattle. The increasing use of the western range as a source brought significant changes to the cattle feeding industry in the Corn Belt. Cattlemen had to decide which of the two available methods to follow in feeding. They could buy cheap western cattle, feed them for a year, then sell them for a small profit per animal and expect to make money on the large volume of business; or they could hope to profit on a low-volume business of better quality cattle that brought a larger profit per head. Not all who chose the latter practice, however, followed it completely, Many did not switch from common cattle to improved beef breeds. Common cattle, though of better quality than those from Texas, were not as efficient in converting corn into beef as stock improved by cross-breeding to purebred bulls. Improved or

22

graded cattle also had better quality meat than the common variety. From the 1840s through the 1870s the Prairie Farmer complained that farmers showed too little interest in improving their beef cattle. However, many saw the problem and tried to do something about it. A Peoria County, Illinois, farmer wanted to know where he "could obtain some cattle that are of better blood than those that have been bellowing over the prairies from time immemorial. Such as a man could take some satisfaction in looking at and feeling of—such as will repay a man for feeding and taking care of."[12]

James N. Brown of Sangamon County, Illinois, provided one answer. The first man of record to bring purebred Shorthorns into the state for breeding purposes, Brown came from the Kentucky bluegrass area in 1834 to found a stock farm; he and his sons were important beef cattle breeders and feeders during the rest of the century. A graduate of Transylvania University and nephew of Benjamin and Elisha Warfield, well-known Shorthorn breeders in Kentucky, Brown was one of the "first to recognize that the best way to get the most profit out of good grass and good corn without robbing the land of its fertility was to stock it with good cattle." In 1855 he had 500 steers, 400 "stall-fed" beeves, and 200 to 400 hogs on his 2,250-acre farm. ("Stall feeding" in contemporary usage did not mean confining cattle to a building but merely to a feedlot.) The farm had 480 acres of corn; 120 acres of wheat, oats, and grass; and almost 1,400 acres of pasture, but Brown's enterprise was the exception before the 1870s. Though several herds like Brown's were in Illinois from the 1840s and in Iowa from the 1850s, only a small number of the common cattle of either state showed much improvement from crossing with purebreds.[13]

Despite the efforts of cattle breeders and agricultural publicists, through local and state fairs, to have "the good blood of our stock . . . widely diffused and encouraged in every condition, thereby elevating the stock of the country generally above mediocrity," most of the fairs reported few improved cattle entries before the middle 1850s. Newspaper accounts of fairs and agricultural society transactions generally praised the cattle exhibited but also noted the relative scarcity and less than desired quality. Except for places such as the Morgan-Sangamon County area in Illinois and Clinton, Scott, Lee, and Story counties in Iowa where purebreds were raised, there was little competition among purebreds for prizes.[14]

Regardless of which procedure a man followed in his

feeding business, he needed capital. The financing of early cattle graziers and feeders was not well documented. It has been argued that the great scarcity of capital resources or their involvement in land purchase prevented the optimum use of much of Illinois and Iowa prairie for cattle feeding. Supposedly, men just did not have the cash with which to finance feeding operations. This lack of capital might have prevented men from starting purebred herds or from buying purebred bulls to improve their stock. Some farmers felt they could not pay "all we have got" just to buy a purebred bull.[15] But feeders such as Funk, Strawn, Alexander, Harris, and Ponting succeeded one way or another in obtaining local cattle on credit or having sufficient cash to buy them in distant areas. Isaac Funk, in effect, borrowed from the neighboring farmers who supplied his feeder cattle by deferring payment until he sold them. After Funk became known to Chicago packers, he occasionally fed fat cattle on contract from a packer who supplied the money for the initial purchase.[16]

Cattle drovers and feeders could also borrow large sums from some local persons of affluence or small amounts from many different people in the neighborhood. In March 1834 Shaw Smith of Paris, Ill., tried to borrow enough cash for a cattle buying trip. He borrowed $60 from one neighbor (but for only a "few weeks"), which would be due before he sold the cattle. He therefore wrote to another man seeking a larger, longer term loan to cover the $60 due and perhaps an additional $60. When some notes came due before the stock had been sold, one man even offered to pay in cattle. Jacob Strawn at first raised wheat and operated a flour mill in order to generate capital with which to buy cattle and land. After that, he frequently borrowed from local merchants and then paid the merchants' bills in St. Louis when he sold the cattle.[17]

Because few men accepted bank notes at face value very far from the place of origin, cattle buyers carried or shipped sizable sums of specie. Alexander and Strawn reportedly carried as much as $20,000 in gold in their saddlebags at one time. Benjamin Harris had some close escapes, but was never robbed of the gold he carried on buying trips. In later years men shipped money by stage or railroad express.[18]

Once a man had surmounted the problem of acquiring cattle, he was faced with how best to feed them for market. Corn and grass were the basic feeds, but cattlemen differed on how to utilize them in the most profitable manner. During the 1840s some insisted that grain was too expensive for winter feeding and, in addition, argued that once cattle had tasted grain they would not want to eat grass. These men believed that grass would be just as effective for growth as grain. If any grain were used, it should have been as a supplement to hay rather than the exclusive feed. Other farmers sought substitutes for hay in root crops such as carrots, rutabagas, mangel-wurzels, and corn fodder. A writer for the U.S. Patent Office estimated that root crops could save from one-quarter to one-half the cost of winter feeding. The Patent Office publications also offered other advice mainly by reporting on English experiments and reprinting articles from the New Genesee Farmer and the Cultivator which, for example, listed the percentage of "flesh" and "fat-forming principles" in certain foods in an attempt to show that foods with the most oil fattened animals the quickest. Thus, farmers should feed such things as sunflower seeds and "all food should be cooked, if possible, and fermented."[19]

Despite the preferences of Patent Office clerks or correspondents, few Illinois and Iowa feeders prepared cooked food for their cattle. The Prairie Farmer contended that availability of corn voided the need for cooked food or root crops. Root crops and cooked food were practical in Europe where little corn was available, but, asked the Prairie Farmer, just because Europeans fed root crops, "is it advisable for us to throw away our corn, and run after turnips and oil cake too?" Rather than follow the suggested variations, Illinois and Iowa feeders stayed with the classic pattern of "stall feeding" developed relatively early in Ohio and Kentucky and brought west. In the stall feeding system men grazed cattle during the summer and fall, then put them in lots during the winter and fed them shocked corn for grain and roughage. As Chicago replaced distant eastern markets and railroad shipping replaced driving cattle to the east, farmers increased the corn ration as market time approached. The proximity of the market and improved transport made it possible to profit from feeding cattle to a better finish than when the long and difficult trip to eastern markets negated the efforts to have a high-quality finish at market time.[20]

Regardless of what periodicals advocated for cattle feeding systems, we do know how some individuals fed cattle before 1850. John T. Alexander gathered local and Missouri cattle for fattening on corn. In the spring and summer he concentrated on raising a large corn crop. In periods of drought, such as 1854, he drove his herd west to the Illinois river bottom to be close to enough corn for winter feeding. As the corn matured, it was cut and shocked in the field where it remained until fed. A few weeks before the pasture gave out in the fall, Alexander gradually introduced a corn ration for the cattle on grass. Then he moved them into feedlots and fed them daily on corn stalks and ears for four to six months. The stalks were thrown from a wagon that followed the same circular path around the lot, so that by the end of the season the "ground was literally floored or paved with cornstalks." The ration averaged about one-half bushel of corn per head per day. Alexander claimed cattle usually gained 200 to 300 pounds in four to six months and doubled in value. He used two feed yards, feeding in alternate lots each day. Swine were placed in the lot not used by the cattle to salvage corn scattered and trampled into the dirt and to scavenge through droppings for kernels that had escaped digestion. Most men raised hogs in conjunction with cattle feeding in order to get full value for the corn fed.[21]

James N. Brown lived several miles east of Alexander in Sangamon County. He specialized in raising and feeding cattle on bluegrass, adding corn only during the last few months before market. On his 2,250-acre farm in January 1856, he had 250 acres of timber, 1,400 acres of cultivated bluegrass and timothy, 480 acres of corn, and 120 acres of wheat and oats. At that time, he had 500 steers of which 300 were going to market in February while the rest remained on hay and grass for the winter. He tried to have enough grass for grazing the whole year. This necessitated having two summer pastures, a winter pasture not used in summer, and harvested hay for lot feeding in March and April when he kept cattle off the pastures. Brown figured it cost him $20 to raise a steer to three years on pasture and then about $18 more to prepare it for the butcher ($4 for grazing and $14 for corn for the last six months). He favored stock improved with Shorthorn blood; he claimed these should yield a profit of $9.50 a head more than average native cattle.[22]

By 1856 Brown advocated marketing finished steers at three years of age instead of four, something that was not common for a good many years. In this system it was best to use calves born in March and April so they would have a full summer on grass. After a winter feed supplement of crushed corn and cob at the rate of two quarts per head per day, the calves came "to grass at one year old fat and sleek, with cost of two dollars per head in corn and cob meal." The calves spent the second summer on grass until mid-December when Brown added a ration of crushed corn at the rate of a peck per day. After eating about twenty bushels of corn in addition to the grass and fodder, the young steers weighed close to 1,000 pounds in May. Grass the third summer in addition to stall feeding one-quarter bushel of corn per day with grass from October to December brought the steers to the feedlot for all the corn they would eat until May. When sold in the late spring, the cattle would average about 1,450 pounds at a cost of $40, which yielded a profit of about $10 per head at the current price of $3.50 per hundredweight.[23]

Benjamin Harris had consulted with Jacob Strawn in 1835 before raising cattle. By December 1836 he had over 100 head feeding on shocked corn at an estimated cost of 10 cents a bushel. About the following April 1, he combined those with others locally purchased for April delivery and turned the whole herd out to graze until July when he started them toward Philadelphia. In the 1850s Harris raised few cattle on his 4,000-acre farm, preferring instead to buy two-year-olds in August and September and stall feed them for the spring market. In the winter of 1853-54 he bought over 65,000 bushels of corn from Sangamon and Mason county farmers in addition to that he raised himself. Cattle were herded on the prairie without shelter in the day and kept in a pen at night; each herd of 100 steers was tended by one hired man. Harris aimed at adding 450 to 600 pounds per animal and expected to make a profit of about $20 per head. In 1856, however, in a contest with a neighbor he produced one lot of cattle averaging 2,377 pounds gross weight per head for a group of 100. In the early 1840s he drove herds to the East and then supplied the St. Louis market, but after 1854 he shipped into Chicago on the Illinois Central Railroad.[24]

Jacob Strawn was one of the first to put corn in shocks for cattle feed, and he was given credit for "setting the example of introducing the present mode of farming," that is, the switch from grazing to feedlot. Strawn, whose great passion in life was "steers—cattle; to drive them, feed them, buy them, sell them," had farms in both Sangamon and Morgan counties, totaling over 16,000 acres in 1854. On much of this land hired hands raised corn. In 1854 he had 2,900 acres in corn and 1,600 head of cattle on the Morgan County farm. Some of the corn land yielded 80 bushels to the acre, but the average was closer to 40. Like Harris, he did not raise calves but preferred to buy cattle for finishing. For many years before the Civil War, he reportedly dominated the St. Louis cattle market, several times allegedly having a corner on it.[25]

All these men had swine that scavenged in the lots either with or after the cattle. Most men believed swine represented a large part of the profit in the operation, and for small-scale feeders swine were assumed to be a decisive element in making the system pay. The ratio of hogs to steers varied from 2 to 1 to 3 to 2.[26]

The above four examples represent common methods of operation for cattlemen, although on a scale larger than typical. Some Illinois and Iowa men fed cattle raised on the farm, though many claimed this was less profitable than buying two- and three-year-olds. It was more profitable, however, than selling the calves to others to feed. The Prairie Farmer waged a vigorous campaign to persuade dairy farmers to raise their calves for beef rather than selling them as young veal or feeders. Some farmers fed more corn, some less; a few tried stable feeding, but most used open lots with a minimum of shelter. Northern Illinois wet areas, east-central Illinois, and frontier Iowa had more grazing on prairie than corn feeding. Some men insisted that rushes in the wet areas were a very succulent forage. Others fed cattle on husked or snapped corn dumped in troughs, rather than corn on the stalk, as a way to save on grain; but they then had to supplement this with hay or straw as a roughage.[27]

Not all men were successful as cattle feeders. The Prairie Farmer through editorials and printed correspondence tried to encourage the inexperienced or less ambitious to develop better practices. Farmers were constantly reminded that they lost money by careless handling of cattle, by a false sense of economy in refusing to spend money on salt or full winter feed, and by not trying to improve quality by crosses with purebred stock.[28]

One much discussed problem of the careless and inexperienced feeder was the "mad itch." To supplement roughage, men turned cattle into picked cornfields or, if they could not afford the labor to shock corn, into fields of standing corn, which were often rented by the acre. Unless plenty of water was available and the cattle had been gradually introduced to dry fodder, they might get dry husks wadded in their third stomach and die of the "mad itch." Men reported the problem fairly often before 1860 and sometimes after. Generally, the advice offered by Prairie Farmer editors and correspondents was to permit cattle in fields only for short periods at first and to have sufficient water available.[29]

Much of the space in livestock columns of the Prairie Farmer during the early years dealt with the specific questions of whether or not it was necessary to provide shelter for cattle on feed. The main argument followed two views: those of humanity to animals and economy of feed. Some argued that leaving cattle in open fields during winter was cruel and sheltering them in barns reduced food costs because they ate less when warmer. Central Illinois men rarely sheltered animals in a barn. To inquiries such as one in 1844 wanting to know "whether the mode in use of yard-feeding cattle . . . is found to combine more advantages than the feeding under shelter," the Prairie Farmer invariably replied that the only right and proper thing to do was to provide shelter, at best in a barn and at worst in an open shed built on the north and west sides of the feedlot. It was "barbarous" and expensive to leave cattle out in storms. The printed correspondence on the subject indicated that many thought feeding cattle in standing corn, timber lots, or on the south and east sides of slopes gave sufficient natural protection. Others constructed hayracks to form enclosures for wind protection. Relatively few mentioned that they stabled cattle. The writers often claimed that cattle sheltered in some way came through the winter in better condition and sold for more in the spring or did not need any extra feeding to bring them back to their prime weight of the previous fall. On the other side of the question, men who did not shelter stock claimed that if the cattle got too comfortable they would not forsake their "nests" to get water or eat at their best pace for gaining weight. Also, it seemed too expensive (to many who did not have the ready capital) to build barns or hire help for the extra work of feeding stabled cattle. Poorer farmers felt that the stock was not worth the expense of shelter or full feed during the winter, but their cattle often sustained a net loss in weight. When these thin animals sold at low prices in the spring, the owners complained that the cattle business did not pay. What cattle needed most was adequate food rather than shelter, and the large-scale operators recognized this in their disregard of more elaborate shelter other than a grove of timber. Despite all efforts to the contrary, barns or even open sheds for stock did not appear to be common before the Civil War.[30]

In addition to feed rations, a cattleman had to cope successfully with market conditions and price fluctuations if he were to operate at a profit. Thus he needed to produce a type and quality of animal suited to the seasonal demands of the particular market open to him. Investing money in better quality beef than the available buyers sought was as likely to be unprofitable as offering poor quality beef for the Christmas trade. The whole industry early shaped itself to the demands of the outside markets in Ohio, New York, and finally Chicago, as well as to local outlets. Illinois cattlemen had first been concerned with Ohio as a marketplace, but limited local markets quickly developed also. O. W. Kellogg pioneered the Peoria-Galena cattle route in 1827 when he took three wagons and a herd from Peoria to the lead mining area. In the early 1830s, Dudley Willits recalled, about once a year young men from Mercer County, Illinois, "would get around fifteen or twenty fat cattle and scare them down" to the St. Louis market for sale at about $8 to $12 a head. By 1851 Morgan and Sangamon counties sent as many as 4,000 head of cattle a week by river to St. Louis.[31]

Southern and central Illinois operators sent herds of cattle weighing from 600 to 1,000 pounds to the East or to New Orleans. Rapidly increasing settlement in northern Illinois and southern Wisconsin created such a demand in the 1840s that many "very sorry animals [that] could never find sale but for want of better" came to the Chicago market from downstate.[32] By 1850 McLean County, Illinois, claimed that it supplied one-fifth of the cattle sold in Chicago. Isaac Funk made Chicago his main outlet after 1835, though he still sold some cattle to New York drovers as late as 1850. Funk even sent his brother Absalom to Chicago to act as his agent over a period of several years. Funk often sold on contract to packers a year in advance at a fixed price and then bought and fed to fill his orders. When driving cattle to Chicago, he moved just enough in one herd to keep a packer busy for one day and spaced his herds one day apart on the road to Chicago.[33]

For several years Iowa cattlemen supplied people heading for California. By the middle 1850s Iowa and Illinois feeders were able to ship over railroads to Chicago and New York, though some men along the rail lines continued to walk cattle into Chicago as late as 1860.[34] New York market reports began listing the number of Illinois cattle offered each week. By 1856, Solon Robinson reported in the New York Tribune that almost the whole New York supply in one week came from Ohio and Illinois and that they were generally three- to five-year-old steers that gave "good evidence of having been where corn is plenty and cheap." Jacob Strawn even sent one of his sons to New York for a short time as his agent for selling the family cattle in the eastern market.[35]

That seemingly endless flow of beef to the East led Robinson to predict that the price was too high in New York and would soon fall. His predictions proved sound but enraged at least one feeder frustrated by a bad year; he replied in common farmer fashion to the workings of the market: "What can we do to stop the mouth or pen of that swaggering and blatant agricultural editor of the New York Tribune . . . who is calling upon us farmers to make up our minds for the low prices? Please tell Solon that we consider him an ass and defy him. . . . We venture to assure all New Yorkers who scribble that we shall sell our cattle to the drovers only at paying prices. Grass fed beef in this quarter is a failure, and we must stall feed what few cattle we turn off, or else sell half fatted stock."[36] Unfortunately, many feeders did the usual thing during a bad crop year when they did not have enough corn and pasture. They sent poorly fattened stock to market and prices fell as predicted.

The representative prices shown in Table B.6 fluctuated from $5 per hundred pounds in 1851 to $1.50 in 1859, and prices were lower on the average toward the end of the period. Prime beef animals for the local fresh meat trade might sell above the top figure given in the newspaper tables. A 1,500-pound average steer would have been worth anywhere from $22.50 to $75 in the Chicago or New York market in the years 1840 to 1860. During the 1850s in Chicago the net weight of the four quarters of beef (dressed weight) from a slaughtered steer averaged between 500 and 700 pounds. Dressed weight probably equaled from 40% to 50% of live-weight.[37] Average gross (live) weights for cattle in the 1840s and 1850s varied from 800 to 1,600 pounds, with weights over 1,200 pounds more frequent in the 1850s. Occasionally, very fat cattle were sent to market, such as the famous herd of 100 cattle averaging 2,377 pounds that Benjamin Harris marketed in 1856. They were four years old and sold at $7 per hundred pounds. A dozen of the herd kept back from the 1856 sale and fed until February 1857 averaged 2,786 pounds when sold. These phenomenal tributes to the art of cattle feeding were reportedly decorated with ribbons and paraded through Chicago streets preceded by a band and followed by one hundred butchers uniformed and mounted. The Prairie Farmer noted that fairs might be giving too much attention to big animals. All things being equal, an 1,800-pound steer was better than an 800-pound animal; but, the Prairie Farmer warned, men were in danger of sacrificing quality for large size.38

Feeders generally sent their fat cattle to market some time after three years of age. In 1847, four- to six-year-old beef was being sold at 600 to 800 pounds net weight after a winter on cheap inferior corn. In the 1840s it cost from $7 to $9 a head to raise cattle to three years of age, while steers could be sold for $15 to $35 a head as feeders. In the 1850s the cost to produce three-year-old cattle increased to an average of $15 to $20 a head, while the market value was about twice the cost. In 1855 an Adams County, Illinois, man paid $30 a head for 1,200-pound steers in October for winter feeding. He anticipated selling them in the spring for about $60 a head after a gain of 300 pounds.39

Bloodline was an important factor bearing on the weight, maturing age, and profitability of cattle. Native cattle matured later, were smaller, and were less economical to feed than purebreds or good grade cattle sired by purebred or improved bulls. James N. Brown figured that he could make a profit of at least $12 a head on 1,800-pound three-year-old Shorthorn grade steers. Grade steers fed the same rations weighed more than natives of the same age; and because there was more and better quality beef on the carcass, they sold for more. However, because grade steers cost more to buy as feeder cattle, many men with limited capital tended to continue raising natives.40

Generous feeding to produce a good fat finished animal was important for profit, but the feed had to be wisely used, for as Solon Robinson reported of a poor quality shipment of Iowa cattle in the New York market, "lots of corn" went "to very little purpose" with bad management.41 Cattlemen had to learn how to get the most gain in flesh on a steer with the least quantity of corn. Two factors were necessary for such a profitable operation: first, men must select steers that had the frame and disposition to readily convert corn into beef, and second, they must provide the proper proportions of corn and grass to keep them gaining weight without wasting feed.

Contract feeding such as that practiced by Funk provided no sure profit. At times contract deliveries upset the market and affected prices and profits. If the price of corn went up because of a poor crop or the grass dried early from drought, drovers were forced to market cattle they had contracted for even if they were in poor condition. This happened in June 1857 when drovers had to take the remainder of their spring contract cattle to market before they were properly fattened, and prices declined. Financial crises, such as that in the fall of 1857, and the timing of the packing season affected Chicago prices too. Packing generally began in October and continued through December, but it varied with the demand, the general state of the economy, and the duration of cold weather.[42]

By the end of the 1850s central Illinois and eastern Iowa cattlemen had shifted their type of operation from one of raising animals to be driven east and fattened in Ohio and Pennsylvania to one of fattening cattle on their own corn and grass for the Chicago and New York markets. The business expanded in volume as the number of Illinois and Iowa cattle sent to market increased, yet the pattern of operations within the region remained the same. Three- to five-year-olds fattened on corn and grass were prepared for either of two types of market. For the butcher market throughout the year cattle were driven or shipped to the city where they were slaughtered and consumed. There was also the seasonal demand of packinghouses that prepared salted or pickled beef during the winter, which could then be shipped to distant places regardless of the season. The increasing number of Illinois and Iowa cattle appearing in both of these markets brought some changes in feeding practices further east. In order to compete with Illinois and Iowa cattle, Ohio feeders began to send animals to market at three and one-half years of age instead of four and one-half.[43] Though Illinois and Iowa men could raise and fatten stock at less cost than Ohio feeders, the Ohio men concentrated on producing fat cattle of better quality in a shorter time, thus eliminating the cost of an extra year in the process. This tended to equalize the difference in production costs. Farther east where costs for fat cattle were higher yet, competition from midwestern stock shipped by railroad caused some New England feeders to give up fattening cattle for the Brighton market serving Boston.[44]

Despite problems in procuring, feeding, and marketing, cattle fattening in Illinois and Iowa was a successful and profitable business by the late 1850s. Events were occurring that would change the patterns of cattle feeding in the next three decades. Until the late 1850s there was insufficient outlet at a good price for large quantities of quality beef cattle. In Chicago, beef not consumed locally was pickled and sold in barrels, a process that reduced all barreled beef to one quality. Differentiation in quality began to be rewarded when men shipped live cattle to New York for the city butcher market, but the greatest incentive for quality beef production came with the development of refrigeration with ice. This made possible the dressed beef trade that offered fresh meat slaughtered in Chicago and shipped in iced refrigerator railroad cars to the urban markets of the nation. In fresh meat sales, better quality brought a higher price. People paid more for fresh meat than for pickled beef, which in turn raised the market price for quality cattle. Farmers could profit by improving their breeds and changing their feeding practices as they did from the 1860s through the 1880s. The entire industry changed. Up to 1860 many Illinois and Iowa cattlemen were ready and able to feed better quality animals, but changes in the market structure were necessary to make this profitable. By the 1860s buyers for the eastern metropolitan markets began to look to the Middle West for their supply of beef; and cattlemen, shippers, and butchers began to look for faster and cheaper ways to get better beef to the cities. This set the stage for the great changes in distribution and marketing that encouraged beef cattle production in Illinois and Iowa.

C H A P T E R 4. CHICAGO AND THE
RISE OF THE PACKING INDUSTRY

In the years after 1865 changes and technological advances contributed to continued growth and expansion of the
beef industry and made Illinois and Iowa leading producers
of corn-fattened cattle and Chicago the world's leading cattle market. The organization and expansion of a central market in Chicago both illustrated and embodied many of these
changes. Several elements combined to bring this market into
being: the new railroad network, the concentration of meat
packers in Chicago, the development of refrigeration facilities, the reorganization of retail meat marketing, and growth
of the export trade in live and dressed beef. As a result
demand increased, and further development of the Illinois
and Iowa farm economy ensued with its advantages for combining the production of cattle and corn.

The necessary first stage in the expansion involved
centralization of marketing facilities in Chicago. Lake
transportation and railroads made this possible by bringing
increasing quantities of agricultural produce to Chicago
for redistribution to eastern markets. As a consequence, a
whole range of commodity handling facilities developed—grain
elevators, stockyards, banks, and a Board of Trade.

Located advantageously for lake transportation by the
Illinois-Michigan Canal Commissioners in 1827, Chicago grew
slowly at first. The canal, which was to bring many benefits,
did not open for 20 years; but militia service during the
Black Hawk War acquainted many men with northern Illinois,
and they returned after the war to establish farms. These
veterans and other new residents turned to Chicago as the
supply center for the region. The federal government constructed harbor facilities and cleared the bar from the
mouth of the Chicago River so that it could serve as a port.
North into Wisconsin and as far south as Danville and Bloom-

ington, Chicago men traded eastern manufactures for western produce. Chicago merchants in turn shipped wheat and pork to balance their accounts with eastern suppliers. In 1838 wheat became the major export factor in this balance of payments and remained so for a decade. By 1842 Chicago shippers sent 586,907 bushels of wheat east, and in five years this had increased more than three times to 1,974,304 bushels. In 1848, the year they organized the Board of Trade, grain dealers had storage space for 700,000 bushels of wheat.[1]

The population of the city grew steadily. In 1840 it had 4,470 residents; in 1850 there were 29,963 residents, and by 1860 the population had risen to 112,172. By 1870 the population was 298,977 and grew to 503,185 in 1880; by 1890 it was 1,099,850. The city itself offered a growing market for midwestern farm produce.[2]

The Great Lakes system provided easy access to the East, and in 1839 a company established regular steamship connections between Chicago and Buffalo. Inland transportation, however, met some serious obstacles. In 1848 completion of the long-promised canal to the Illinois River opened a water route to the area southwest of the city. In the same year newly constructed plank roads offered another way through the wet prairies of the low watershed between Lake Michigan and the Illinois River. Some of these extended almost 60 miles from the city, but they were relatively unimportant in comparison to the railroads and water routes.[3] Though both the canal and plank roads were important in transporting grain, they also aided the marketing of meat products, the other mainstay in agricultural commerce.

Pork packing was the first animal product business of any magnitude in Chicago and the Middle West. Cincinnati held the first large concentration of pork packers, but practically every river town in the Ohio and upper Mississippi river valleys had a packing establishment at some time. For example, during the six-month season ending in March 1848 six Chicago firms packed 19,666 hogs, and 14 towns on the Illinois River and 9 Illinois and Iowa towns on the Mississippi River packed from 1,500 to 35,000 hogs each. Though pork was their major product, some packers also slaughtered beef in Chicago and in the larger towns such as Alton, Peoria, Rock Island, and St. Louis.[4]

George W. Dole (pioneer merchant) started the first regular packing business in Chicago in 1832. In the first season he slaughtered and packed 152 head of cattle and 338 hogs.

Soon after that Gurdon Saltonstall Hubbard (fur trader and
frontier merchant) entered the business, and by 1836 his
firm packed 6,000 hogs in one season. The financial crisis
of 1836 dealt a temporary blow to the economy by halting
work on the Illinois-Michigan Canal. As local demand for
pork and beef temporarily declined, men tried to market
their barreled meat in the East. In this trade pork products
maintained a considerable lead over beef in the 1830s, but
by 1843 Chicago beef shipments totaled 10,380 barrels compared to 11,112 barrels of pork. The firm of Wadsworth,
Dyer, and Chapin (the leading beef packers in the 1840s) at
times slaughtered 100 head of cattle a day and packed beef
for the U.S. Navy and the British government. Thomas Dyer,
partner in the firm, so systematized the disassembly process
that by 1847 the plant could handle 45 animals an hour. Coincident with this development of beef packing, tanners,
coopers, and fat processors established businesses.[5] Most
cattle driven to the Chicago market originally consisted of
grass-fed animals, but by the late 1840s Chicago packinghouses began to offer an alternative market for Illinois
corn-fed beef formerly driven overland to eastern markets.[6]

Before the development of artificial refrigeration,
beef packing was a seasonal operation limited to the period
of cold weather. In Chicago the packing season generally
opened in late October or early November and could last until early March, depending upon demand and the state of the
weather. The operations of Wadsworth, Dyer, and Chapin offer
an example of large-scale beef packing in the late 1840s.
During the 60-day season in the fall of 1848, the company
planned to slaughter over 3,500 cattle. On peak days 75 to
80 men slaughtered and packed from 120 to 180 cattle a day.
Men cut carcasses into eight-pound pieces, put the meat
through a three-step pickling process, packed 38 pieces in
salt into a tierce (a cask between barrel and hogshead size)
and then shipped the tierces to New York where they sold at
$18 each for the English market. After the better pieces of
meat had been barreled, the workmen cut, salted, and dried
the necks and shanks for sale the next spring in the West
Indies as jerked beef. The refuse fat produced 40 barrels of
tallow a day, which sold for 7.5 cents a pound. Wadsworth,
Dyer, and Chapin sold tongues, hoofs, horns, and gall, but
did not use the liver, lungs, bones, or blood.[7]

To supply their packinghouse, Wadsworth, Dyer, and
Chapin bought cattle from as far south as De Witt County,

Illinois, and the upper Wabash valley. Isaac Funk was one of their major suppliers. He furnished 1,200 head in October 1848, 1,650 head in November 1849, and 1,000 head in November 1850. Most dressed out at between 600 and 750 pounds and were purchased for about $3 to $3.50 per hundred pounds live-weight. Sometimes the packers put up the capital for Funk to buy, feed, and drive cattle to Chicago on contract.[8]

As early as 1849 the Prairie Farmer declared that "Chicago is at present the first point in the United States in slaughtering and putting up of beef. There are some two or three establishments engaged in this business here, who have no rivals at any other place in the United States—so we are informed, and so we believe."[9] In 1850 the eight establishments listed by the Chicago Tribune expected to pack 27,500 cattle, which at a conservative estimate represented an investment of almost $0.5 million in animals alone. Despite the Prairie Farmer's boast Chicago's total quantity of packing was not that remarkable. Beardstown, Peoria, and Quincy did a large pork packing business and also packed some beef.[10]

The further development of Chicago as a meat packing center depended upon an improved transportation system. Canal and lake routes closed during the winter, and the vagaries of the weather made driving livestock to the city a risky venture. In the spring, roads were almost impassable. The railroads overcame the handicap by providing all-weather connections with the hinterland. With the Chicago market readily accessible by rail, prices there became the standard for other western markets. By the early 1850s centralization of packing facilities had taken place to the extent that for a century the history of the meat processing industry in Illinois was largely the history of meat packing in Chicago.

In the late 1840s and early 1850s men around the state began to revive earlier plans for internal improvements. This time there was a considerably larger economic base and immediate prospect of revenue. The Galena and Chicago Union Railroad began pushing out to the wheat country west of Chicago in 1847 as a direct result of agitation by the business community to surmount the transport difficulties in that direction. Petitions to the legislature asking the state to build the railroad had drawn little support. However, Chicago interests raised the capital and began work; by 1849 trains ran to a point 21 miles west of the city. Six years later the railroad ran to Freeport, which was 122 miles northwest, straight west to Fulton on the Mississippi, and to Beloit, Wisconsin.[11]

Other railroad projects followed. Construction of the Chicago, Alton, and St. Louis Railroad began in 1847, and by 1854 it offered through service to St. Louis. The great Chicago, Burlington, and Quincy system was launched in 1849; reached the Mississippi across from Burlington, Iowa, by 1856; and was extended to Ottumwa by 1859. By 1870 a direct line ran through southern Iowa from Council Bluffs to Burlington and on to Chicago. The Chicago and Northwestern won the railroad race across Iowa to Council Bluffs in 1867 by taking over the Galena and Chicago Union and consolidating and extending Iowa short lines. The Chicago, Rock Island, and Pacific linked Chicago to the Mississippi at Rock Island in 1854, and by 1869 it too had connections with Council Bluffs. The Illinois Central Railroad, started in 1851 as a north-south line down the center of the state, had 700 miles of track in operation between Chicago and Cairo and Cairo and Dunleith in the northwest corner of Illinois by 1858. It too then crossed the river by 1870 and had built westward to Sioux City in western Iowa.[12]

In 1860 six major rail systems extended south and west of Chicago, tying Illinois and Iowa closer to the city and encouraging settlement of the prairies. (See Map A.5.) Illinois had 2,790 miles of railroad in operation or 1 mile of railroad to every 19.8 square miles of land. Iowa, at the same time, had only 655 miles of railroad or 1 mile for each 84 square miles. By 1900 both states were thoroughly crisscrossed by railroads. (See Table B.7.) However, railroad building continued at a brisk pace, so that in 1880 Illinois had 7,851 miles of track and Iowa had 5,400 miles. By 1900 the railroad systems of Illinois and Iowa had grown to the point where they had 11,058 and 9,391 miles of track respectively. (See Map. A.6.) Railroad development tied Illinois and Iowa together as one economic tributary to Chicago where the railroads converged. By 1900 at least 19 railroads brought cattle into the Chicago market.[13]

This railroad building had immediate and far-reaching impact on Chicago and the cattle business. By making beef more readily available to Chicago and the eastern markets, the railroads stimulated both consumption and production. As one Iowa farmer wrote in 1856 after the Burlington Railroad had penetrated Wapello County, "many of us are turning our attention to feeding cattle." The coming of the railroad also led him to ask the Prairie Farmer to print wholesale cattle prices and to publish information about shipping rates and routes to New York.[14] It cost about $15 a head to ship live cattle from northern Illinois to New York and about $20 a head to ship cattle from Burlington, Iowa, to New York; over two weeks was required for the trip from Burlington.

The Prairie Farmer calculated that people in New York consumed 185,574 beef cattle in 1855, while in Philadelphia and Baltimore the population consumed at least 100,000 head. Half of the total beef consumed in these cities came from Ohio and the West. Illinois alone sent 17,000 cattle east by rail in 1855.[15] One prominent cattle raiser estimated that the coming of the railroad increased cattle prices in central Illinois by 10%. The railroad also encouraged rapid settlement of lands in eastern Illinois and in Iowa. Prairie-county populations in east-central Illinois increased from 150 to 600% from 1850 to 1860, and the acres under cultivation in the prairie region went from 74,000 acres in 1850 to 2,902,000 in 1860.[16] After the Illinois Central Railroad began operation, Effingham County, 200 miles south of Chicago, "increased rapidly in population, and many of the farmers are now driving a very respectable business." Similarly, in Iowa, farm products brought "a good price" in Cass County once the railroad arrived in 1869. Seeking the same benefits, Jefferson County Iowans organized to secure a railroad line through their area, which would make "things begin to look up." At that time, corn in Jefferson County (without a railroad) sold for 15 cents a bushel less than in Cass County, which was served by a railroad.[17]

Initially, however, railroads seemed to have an adverse effect upon the number of cattle packed in Chicago. Compared to the 27,500 packed in 1850, during the short season of fall 1852 packers processed a little less than 22,000 cattle. There had been a heavy summer demand in the East and the railroads permitted shipment of live animals to New York and other eastern markets. The shipment of beef animals to eastern markets increased from an estimated 2,657 in 1853 to 42,638 head in 1858. In 1858 Chicago firms packed 34,675 head of beef. By 1860, 97,474 animals were shipped east, enough to have supplied 43% of the number received in New York City that year. Of the remainder of the Chicago receipts of 177,101 cattle, 34,623 animals were packed for shipment and 42,074 were consumed within the city. Chicago Board of Trade figures indicated that about 20,000 cattle were driven to Chicago in 1860; the rest arrived on the railroads. By 1864, under the influence of the Civil War, receipts had climbed to over 300,000 with over 70,000 packed for markets outside Chicago.[18]

As railroads brought more cattle and hogs into Chicago, local boosters proclaimed its preeminence as the packing center of the nation. Chicago establishments packed half a million hogs in 1862, surpassing Cincinnati as a pork packing center. At the same time, receipts of cattle for slaughter and packing increased. A Chicagoan proclaimed the city the "great meat manufactory of the world."[19]

But even as Chicago made this boast, in 1865 the number of cattle packed started to decline. Live cattle receipts and pork packing continued to grow, but the beef pack of 92,459 animals in 1865 was the high point for many years. In 1870 Chicago packers slaughtered only 11,963 cattle. The decline occurred because some packers had moved west to Kansas City to utilize cheaper range cattle for barreled beef where quality differentiation was not important. In addition, eastern buyers paid to ship numbers of live cattle to the East for slaughter there.[20]

The Civil War was a boon to Chicago as well as to Illinois and Iowa beef producers by increasing demand at a time when railroads could deliver beef to the city. The war and the railroads also brought centralization of market facilities as increasing receipts of livestock created chaos in handling transactions between several markets in the city. Relief came with the formation of the Union Stock Yards in 1865.

Before 1865 Chicago had several small stockyards located in various parts of the city. In the early days drovers immediately sold the few cattle they brought in. There was no need to pen cattle while the drover sought a buyer. But as the trade increased, there was need for a place to keep live animals before they were sold. The slaughter houses had some pens, but private yards soon superseded them. In the late 1830s Willard Myrick opened one of the first of these yards as an adjunct to his tavern where drovers stayed while seeking a sale. In 1848 the Bull's Head Yards opened near the southwestern plank road. Entrepreneurs opened similar yards around other taverns, so that a number of possible marketplaces were available. With the advent of the railroad, yards not advantageously located fell into disuse, so the duplication of facilities declined. In the late 1850s, however, the

41

existence of several yards still required buyers and sellers alike to ride the circuit to close an advantageous deal. Cattle prices could vary as much as 50 cents per hundred pounds between yards on the same day.[21]

During the Civil War, representatives of the packers and the railroads began negotiations to end the inefficient proliferation of stockyards. Packers had to have buyers at several different places, and railroads had difficulty getting shipments from the yards of a western railroad to the yards of an eastern one. Buyers had attempted to meet the problem by designating Tuesday and Friday only as market days for cattle; and in 1856 the junction railroad linked all the railroads entering Chicago, but that was no solution to the problem. At a meeting in 1864 the Packers' Association called for a central stockyard. The railroad managers accepted the idea and applied for a charter from the state legislature.

In June 1865 the Union Stock Yard and Transit Company of Chicago began construction of the Union Stock Yards and opened the gates for business December 25. Four representatives of the packers and five for the railroads composed the first board of directors. The company had a capital stock of $1 million, all but $75,000 held by nine railroads. The yard covered 345 acres (only 100 acres of pens at first); had two artesian wells, a drovers' hotel, and a bank; and could care for 100,000 head of stock. Four separate unloading areas served the five western and four eastern railroads that entered the yards.[22] The railroads controlled the yards until 1894 when their representatives resigned after losing control to new interests. As a result of a complicated corporate shuffle typical of that decade, Armour and Company (one of the major meat packers) and Frederick Prince (Boston banker and broker) took control. The packers had wanted a share in the profits and management for some time. They finally threatened to move their operations and build new yards of their own, at which point some of the old stockholders and railroads gave in and allowed a corporate reorganization.[23]

Farmers complained in the 1870s and 1880s that the Stock Yards made an exorbitant profit from the yardage fee of 25 cents a head for cattle and feed charges of $1 a bushel for corn and $20 to $30 a ton for hay. Yard prices ranged from 50 to 300% above the market price outside the gates. In 1877 an Illinois legislative investigation found that a basket of corn called a bushel by the yards did not in fact weigh 70

42

pounds but only 55. The current market price for a 70-pound bushel was 43 cents, while the yards charged $1. A bill to have the state take over the management received little support in the legislature, but the State Railroad Commission obtained supervisory powers allowing it to set maximum limits on rates for feed and handling. A later bill in 1887 to regulate the commission and feed charges of the yards did not pass.[24]

Despite complaints the yards immediately proved their usefulness, and as years went by they were improved and enlarged. By 1887 the capacity was 20,000 cattle, 150,000 hogs, 10,000 sheep, and 1,500 horses. By 1900 the yards covered 500 acres and had 250 miles of railroad track, 130,000 pens, and a capacity for 75,000 cattle, 80,000 sheep, 300,000 hogs, and 6,000 horses. Over 200 commission firms had offices there.[25]

With minor fluctuations the number of cattle handled each year continued to increase until 1892 when it reached the high point for the nineteenth century: 3,571,796 cattle and 197,576 calves. In 1900 the yards handled only 2,729,046 cattle and 136,310 calves. (See Table B.8.)[26] The reduction in receipts after 1892 came partly because of the decentralization of the meat packers to Kansas City, Omaha, St. Paul, and Iowa cities and partly because of a series of years of low prices during which breeding herds had been reduced and some Illinois and Iowa men gave up cattle feeding.

In addition to the handling services the yards provided space for commission firms and financial institutions. The Union Stock Yards National Bank, established in 1869, became the National Live Stock Bank in 1888 and had assets of over $9 million in 1900. The Drovers National Bank, organized in 1883, had assets over $5 million by 1900. There was also the Chicago Cattle Loan Company, with much the same directorate as the National Live Stock Bank which had several Union Stock Yards men and packers' representatives on its board.[27]

By 1870 centralization of marketing and packing facilities tied terminal and transshipping railroads, commission houses, financial institutions, and packing companies together into a "packing town" built around the Union Stock Yards on the south side of Chicago. However, with all this change in handling cattle, there had been none in beef processing procedures for a growing market. Quality cattle for the eastern market were sent live, a costly and inefficient way of handling a large volume of business, which tended to

43

make Chicago only a transshipping point.

The Chicago market structure and processors could handle more cattle than the barreled beef market demanded. Chicago was prepared to handle range cattle before they were readily available after the Civil War; and yet when shipments increased the volume of the market in Chicago, cattlemen found that the quality of range beef offered more meat for barreling than for the live trade to New York. Range cattle were thus no answer for the rising urban demand for quality fresh meat. However, the combination of the ability of Illinois and Iowa feeders to produce quality beef cattle and the centralization of marketing, processing, and shipping facilities of Chicago with a new technique for handling fresh beef brought renewed prosperity to the industry. Quality beef could be slaughtered in Chicago and sent to the East in refrigerated railroad cars, where it could be marketed as fresh beef without the expense of shipping all the offal, the liability in shipping live cattle. The development of this dressed beef trade accounted for much of the growth of the beef cattle industry in the last half of the nineteenth century.

C H A P T E R 5. DRESSED BEEF AND
THE EXPANSION OF THE PACKING INDUSTRY

The demands of the new urban market, created by the expansion of the railroads, gave rise to the dressed beef industry. Consumer demand for processed agricultural goods in
urban areas made it possible for those supplying the products to build national marketing organizations and to concentrate the national trade of any line in the hands of a
few firms. The processors of consumer goods such as flour,
tobacco, and fresh meat were the first big businesses created by the rise of the urban market.[1] A rising demand existed for fresh meat in the East as the cities outgrew their
local supplies after the Civil War. A supply of quality beef
was potentially available in Illinois and Iowa feedlots if
there were sufficient price rewards. Chicago market facilities could handle an increased volume of cattle. Both the
producers and the processors were prepared for the change
when the possibilities of the lucrative urban market encouraged men to combine the railroads with refrigeration to ship
fresh beef slaughtered in Chicago to the eastern markets.

The introduction of effective means of refrigeration
prevented the early spoilage of fresh meat and lengthened
the period of time after slaughter and the distance from
point of slaughter that fresh beef could be consumed. The
basic principles of refrigeration had been understood and applied, with ice compartments in meat cooling rooms, before
the Civil War; but the packing industry was slow to use cooling systems in marketing fresh meat in conjunction with railroads. There had been some experiments with ice refrigeration
in railroad cars, however, and by 1869 George Hammond, a De-
troit packer, sent fresh beef to Boston in a refrigerator
car. Packers and shippers began to appreciate the possibilities of refrigeration in the early 1870s and installed chill
rooms in the packinghouses. Most of this early refrigeration

resulted from air circulating around ice or an ice-salt mixture. Not until the 1890s did mechanical refrigeration of the ammonia-compression type come into common use in the meat processing industry.[2]

Hammond continued his experimentation with refrigerator cars and started sending small shipments regularly to Boston in 1872. Nofsinger and Company, Kansas City packers, sent two carloads of dressed beef to Philadelphia in 1875 as well as in the next few years, but did not continue partly because railroads raised the rates to discriminate against the product. A Chicago group gave a hint of things to come when in 1875 they tried a scheme to undercut butchers supposedly planning to raise meat prices in Philadelphia during the centennial exhibition. They sent dressed beef to Philadelphia in ice-packed cars at a temperature below 35 F and sold directly to consumers at a reduction of 34% from the current butcher price. The Prairie Farmer suggested that such direct retailing be done on a large scale in New York City.[3]

Gustavus Swift (a butcher and retailer from New England) led the way in employing the new technology of refrigeration to concentrate beef slaughtering and processing in Chicago and as a result changed the marketing system for fresh beef. Swift had moved west first to Albany, N.Y., and then in 1875 to Chicago in order to eliminate middlemen in his marketing operations. At first he shipped live cattle east to his Massachusetts partners; but when he saw some of the experiments in shipping dressed beef, he decided that the competitive advantage lay with the man who perfected the method. Swift set about creating the dressed beef processing and distribution system that became the standard of the packing industry well into the twentieth century. In the winter of 1876-77 Swift shipped dressed beef in boxcars without refrigeration; this was unsatisfactory, so he experimented with refrigerated cars. When the major railroads refused to build refrigerator cars, Swift ordered them from independent builders and by 1879 he owned 100.[4] Once Swift had created a demand for dressed beef by underselling local butchers through his own retail outlets, he borrowed all the money he could, as well as reinvesting his profits, for expansion to match demand. Swift erected a Chicago slaughterhouse in 1880 and opened houses in Kansas City in 1888; Omaha in 1890; East St. Louis in 1892; St. Joseph, Mo., in 1896; and St. Paul, Minn., in 1898. By 1899 Swift had 193 branch houses distributing his products to retailers across the country.

Incorporated with a capital of $300,000 in 1885, Swift and
Company had a capital of $25 million and annual sales of
$160 million by 1903.[5]

The dressed beef trade became important almost immediately. (See Table B.9.) By 1882 the <u>Prairie Farmer</u> reported
that Swift's shipments had caused a "sensation" in eastern
markets. The next year, realizing the potential of the
dressed beef trade, Armour and Company and Nelson Morris,
leading Chicago pork packers, joined Swift in shipping
dressed beef. At first, New England was the major market; as
it developed, there was an immediate decline in the number
of live cattle shipped from Chicago as well as from St. Louis
to New England. (See Table B.10.) But soon packers shipped
to New York, Philadelphia, and Baltimore. Indeed, from January 1881 to November 1, 1884, almost 89% of the dressed
beef received in those cities and all of New England came
from Chicago. By 1883 domestic and export demand was so
great that even with record cattle receipts during September
prices did not decline.[6]

There was a bitter 15-year struggle before dressed beef
was fully accepted by consumers, butchers, and railroads.
Consumers distrusted the quality of meat shipped hundreds of
miles after being slaughtered, but dressed beef was suffi-
ciently cheaper than local butcher stock that people were
usually willing to try it. Swift had a much harder time convincing butchers to sell Chicago-dressed beef. Local butchers controlled retail meat distribution in the nation. They
sometimes obtained their meat from large city packinghouses
or killed cattle on the farms where they bought them; in
most cases they bought cattle at a stockyard and dressed
them in a local slaughterhouse or their own shop. Butchers
and owners of local slaughterhouses, who saw the threat to
their business, fought back and had the support of railroads
that had an investment in the live cattle trade.[7]

A stockcar of live cattle produced fewer pounds of salable meat than were carried in a refrigerator car of dressed
beef. If railroads charged the same rate per pound for both
live cattle and dressed beef, the processed meat could be
shipped at considerably lower cost per salable pound. Retailers of dressed beef could then afford to undersell local
butchers; if enough butchers sold dressed beef, there would
be less demand for live cattle shipments. The shippers saw
this clearly and prevailed upon the sympathetic railroads to
set rates per car or ton-mile for dressed beef considerably

higher than for live cattle. This reduced the price advantage for dressed beef to the benefit of cattle brokers, stockyards operators, and the railroads with their investment in cattle cars. New York butchers joined the group pushing for higher rates in order to reduce or eliminate competition from dressed beef. The Prairie Farmer thought the railroad interests had organized the butchers but assured Cornelius Vanderbilt (of the New York Central Railroad) that "manipulating stocks is child's play compared with this new venture" to ruin the dressed beef trade.[8]

The Vanderbilt railroads and the Pennsylvania system, with large investments in cattle cars and stockyards, refused to provide refrigerator cars or to haul them at advantageous rates. As a result, packers not only built their own but turned to the Grand Trunk and the Baltimore and Ohio railroads to haul them since those lines had very little live cattle business. Even so, the other railroads and live cattle shippers did succeed through rate pools in keeping published freight rates for dressed beef usually at least 50% above those for live cattle from 1879 to 1892, even during the rate wars. In 1883 Chicago firms could place choice cuts of beef in New York for 11 cents a pound. The New York butchers claimed they had to pay 12 cents a pound for live cattle and charge 25 to 35 cents for choice cuts in order to pay for labor, waste, and loss on less desirable cuts and still make a profit. If butchers bought from one of the packer branch houses, they could have the cuts they wanted at 12 to 13 cents a pound, sell them for 18 to 20 cents, and still make a profit. At that time live cattle freight rates to New York were 40 cents a hundred pounds and dressed beef rates were 64 cents.[9]

Despite shipper protests and resolutions by such groups as the National Cattle Growers' Association supporting lower rates for dressed beef, the railroads refused to reduce charges to the same level as for live cattle. Only the Baltimore and Ohio and the Grand Trunk railroads seemed inclined to bolt railroad rate pools and thus caused a rate war in 1887 and 1888. Over the last quarter of the century, the level of rates on all kinds of beef generally declined, but there was no equality for live and dressed beef. However, by 1890 the Chicago packers had sufficient organization and economic power to influence railroad rates to their advantage.[10]

Local butchers, especially wholesalers, were most ve-

hement against dressed beef and combined to fight the packers. They organized the Butchers' National Protective Association in 1886 and attacked dressed beef on the grounds
that it was unwholesome and that only the locally killed
product was sure to be sanitary. In some areas they enlisted
the support of local trade unions to boycott butchers who
retailed Chicago dressed beef. The packers responded by establishing branch wholesale houses in cities, opening retail
outlets, and selling it as low as 3 cents a pound. When local butchers agreed to sell dressed beef from the branch
house, the packers closed their retail stores. Packers also
established the peddler car, a refrigerator car filled at a
branch outlet to deliver small amounts of dressed beef and
other perishable goods to little towns along the railroad
line. This technique created an almost limitless national
market for fresh beef, mutton, and pork and seriously threatened local butchers in every town served by a railroad.[11]
Some states adopted legislation forbidding price cutting and
other packer retail practices, but U.S. district courts, the
Supreme Court affirming, struck down these state laws as re-
straints of trade. By 1894 all the obstructionist efforts
had failed and the Butchers' Advocate acknowledged defeat.[12]

The hermetically sealed tin can provided another way
that Illinois and Iowa corn-fattened cattle could be packed
and shipped to domestic and foreign markets. The tin can had
been known and used since the Napoleonic wars, but not until
the Civil War and the use of canned condensed milk did it
become popularly accepted in the United States. Soon after
the war Chicago packers began to investigate canning as a
replacement for packing pickled beef in barrels. William J.
Wilson (a leading packer) began canning beef in the early
1870s; Libby, McNeil, and Libby began in 1875. After a court
ruling invalidated patent claims on the canning process, all
packers handled canned meat. A regular canner grade of cattle
came to be recognized in the market. This was mostly western
but also included some lightweight Corn Belt cattle averaging about 950 pounds each. Two canning processes developed.
One consisted of placing boiled compressed meat in square or
pyramidal cans, which were then reboiled and sealed after
cooling. The other consisted of putting raw meat into round
cans, which were cooked and sealed before cooling. By 1884 a
Bureau of Animal Industry agent reported that half the canned
meat of Chicago packers went to Great Britain.[13]

In addition to the domestic trade, exports of livestock,

49

dressed beef, and canned beef had an important influence on
the industry. The foreign trade took up the slack as American beef production increased, and this had a definite influence on market conditions. There had always been some export of cattle and beef products from the United States, but
it did not increase significantly until after the Civil
War.14

Most exported live cattle went to England from the late
1870s to the late 1880s, while through the end of the century the majority of dressed beef went to England. The English were faced with a serious problem, as the urban population grew much faster than local agricultural production. At
the same time, the United States had a growing surplus of
cattle, and livestock reporters noted with favor the growing
foreign demand that buoyed the market.15

Nelson Morris (the first of the major packers in Chicago to cater to the European trade in beef) shipped some
live cattle to England in 1868, but not on a large scale.
Next, a Scottish firm imported some American cattle in 1873
and 1874, but did not continue. Finally, Timothy Eastman (a
New York cattle dealer) revived live shipments to England in
1876; and from then until 1885 he was the major exporter of
live cattle. During July, August, and September 1878, for
example, England imported 22,129 live cattle from the United
States.16 Continental cattle feeders even went to the extent
of importing feeder cattle for fattening and resale in England. A Schleswig-Holstein group paid as high as $137 a head
for 1,100- to 1,200-pound steers delivered in Hamburg to be
fattened for German and English markets. This prompted the
Prairie Farmer to predict the development of a whole new
market for American feeder stock in Europe. However, in the
late 1870s Britain restricted cattle imports from the Con-
tinent for fear of spreading cattle diseases, and pressure
from English feeders and fear of pleuropneumonia in American
cattle brought restrictions requiring them to be slaughtered
on the wharf at the port of entry. The Prairie Farmer immediately called for federal laws to control contagious diseases so that beef cattle could be certified disease-free.
Chicago prices fell in January 1879 when the new British restrictions went into effect, but these affected the live
cattle trade only; dressed beef shipments increased.17

Despite a temporary decline, shipments of live cattle
to England continued. The rule requiring immediate slaughter
of imported live cattle remained in effect until the early

twentieth century, even after American beef processors and feeders had succeeded in obtaining government action on disease control and inspection programs for livestock. By 1900 the United States exported 396,977 live cattle, the majority of which went to England and almost two-thirds of which passed through the Chicago market.[18]

Dressed beef had the same advantages over live cattle in the export trade as it had in the domestic market in that more salable pounds of dressed meat than live cattle could be shipped in the same space. The _Prairie Farmer_ and other farm periodicals, as well as Illinois and Iowa fat-cattle producers, were interested in the foreign market. Articles reported opportunities in the English market, London prices, and problems posed by various types of embargoes by foreign governments for protection from American pleuropneumonia. Papers also followed reports from English newspapers on beef shipments originating in the Middle West. Illinois beef carcasses arrived in Glasgow, Scotland, as early as June 1876, having been slaughtered in New York just prior to shipment. A Scottish newspaper reported in the early fall that sales had averaged 150 carcasses a week.[19] During July, August, and September 1878 England imported 59,512 quarters of beef from the United States, while during the first six months of the year 165,422 quarters of beef entered that country. All this dressed beef came iced, but refrigerator-equipped ships were being planned to reduce the load of ice and increase the amount of beef carried. Liverpool received 2,120 live cattle and 4,280 quarters of beef in one week in October 1878. The magnitude of shipments grew so much from 1875 to 1878 that shipping lines converted passenger ships to beef carriers.[20]

By 1900 the United States had exported 329,078,609 pounds of dressed beef, mostly to England. Table B.11 compares 1900 exports in volume and value with selected other products. The value of live cattle slightly exceeded that of dressed beef, and the combined total value of beef made up 60% of the value of the total meat exported, the third largest item in value of American exports.[21] (See Fig. B.1.)

The great expansion of the market through technological and marketing changes in the dressed beef trade created large, vertically organized corporations based in Chicago. These corporations controlled, for the most part, both the wholesale and retail domestic markets for beef products. Known as the "Big Five," Armour and Company, Patrick Cudahy,

Nelson Morris, Swift and Company, and Wilson and Company frequently acted together in buying cattle in the Union Stock Yards and in apportioning retail markets among themselves.

Occasionally the Big Five met to set prices and attempt to eliminate some of the worst aspects of competition. With high capital demands and rapidly expanding business, they were pressed for cash and took every means to reduce overhead. The first pool organized in 1885 included Swift, Morris, Armour, Samuel Allerton (farmer, stockyard owner, and packer), and George Hammond and concerned only shipments of meats on a pro rata basis. The second known pool operated from 1893 to 1896 when Henry Veeder (a lawyer for the packers) acted as pool secretary and compiled reports of shipments to the major eastern market areas, apportioned them among the pool, and assessed fines for overshipments. Representatives of the packers met every Tuesday afternoon to discuss operations. The pool broke down in 1896 and 1897 when Schwarzchild and Sulzberger (a New York based wholesaler and packer) flooded the market; however, that firm joined the pool in 1898, and the packers set price margins and shipment quotas through the end of the century. The pool even set prices by ascertaining the net carcass cost of average operations for each packer and then establishing a uniform minimum cost system for the industry regardless of the price of livestock. By basing the wholesale price of meat on an artificially high cost, they all made a profit.22 A Federal Trade Commission investigation revealed this only after the statute of limitations had expired; and when they were under investigation, the packers denied operating any pools in the early twentieth century. Contemporary rural newspapers and a Senate committee attacked the packers in the 1890s for being against the cattle feeders' interests. Packers were criticized for driving local butchers out of business by selling beef below cost. Once the local butchers closed their doors or agreed to be distributors for the packers, retail meat prices rose and wholesale prices to feeders supposedly declined. Packers were also accused of using range cattle as a "club with which to beat down the price of natives [cattle raised in the Middle West] and compel the farmer to compete with the range." Some farmers became discouraged and let the quality of their cattle deteriorate to range level. The only remedy offered by the press was for farmers to breed and feed prime cattle for the eastern or export market. Live shipments to the East had increased in 1890 and 1891 so newspapers urged farmers to "breed past Chicago."[23]

Some packers even went into cattle feeding themselves. As early as 1878 Nelson Morris kept cattle in Peoria, Illinois, which were fattened on distillery mash. Morris even helped finance construction of the Great Western Distillery, said to be the largest in the world at the time. Cattle chained to feed bunks received wet mash with hay for balance. In 1880, 16,000 cattle were fed on mash in Peoria. The number fluctuated and rose as high as 28,000 in 1893 but declined to less than half that by 1895. Morris also had western ranches and a large farm in Indiana where he fattened range cattle. He turned out as many as 75,000 head a year from his feedlots.[24]

By the early twentieth century when the federal government investigated the packers, they were large corporations. The Big Five killed 70% of all livestock and 82% of cattle slaughtered for interstate commerce. All were very much interested in meat substitutes such as dairy products, fish, poultry, eggs, and vegetable oils. Swift was the greatest butter distributor in the country by 1916, and the five controlled at least half the interstate commerce in poultry, eggs, and cheese through their wholesale branch houses and retail outlets. They refined 31% of American cottonseed oil in 1916, giving them a big stake in the oil-cake business as it became a more popular cattle feed supplement. They were also major factors in the hide and leather industry. The Big Five owned 91% of all the refrigerator cars equipped to ship dressed beef on the railroads, and they controlled the major stockyards, terminal railroads, market papers, and livestock banks. Packer men sat on the boards of 3 banks in Boston; 9 in New York; 25 in Chicago; 5 in St. Joseph, Mo.; 3 in Kansas City, Omaha, and Wichita; 2 in San Francisco and St. Paul, Minn.; and 1 each in East St. Louis, Denver, Fort Worth, Oklahoma City, Portland, Oregon, and Sioux City, Iowa.[25] Much of the magnitude of other interests of packers became apparent after 1900, but the initial basis for this growth came from the rise of the nineteenth-century midwestern beef and pork processing industry.

Chicago remained the meat marketing and processing center of the nation in 1900, though the packing industry began to decentralize. By the end of the century the leading packing centers and number of cattle slaughtered were: Chicago

(1,794,397), Kansas City (1,092,804), Omaha (516,669), East St. Louis (484,008), and St. Joseph (228,977). Lesser centers were Des Moines, Sioux City, St. Paul, and Fort Worth. The packers had so revolutionized the meat trade that butchers in some cities claimed that "not a retail butcher has made a fair profit and a living in the last ten years" because of packer branch houses.[26] The packers had also affected the raising and feeding of cattle in Illinois and Iowa.

C H A P T E R 6. STRUCTURAL CHANGE IN ILLINOIS-IOWA
CATTLE FEEDING TO 1900

Expansion of the packing industry went hand in hand
with changes in beef production. Demand for beef increased
because of growing population, improved transportation, and
better preservation of meat. The vision of profits, touted
by word of mouth, newspapers, and books, attracted men and
capital to the Great Plains to establish or enlarge cattle
ranches. The increasing range cattle production, in turn,
encouraged changes in midwestern feeding operations. Illinois and Iowa producers who did not fatten range cattle were
required to improve the quality of animals they raised and
at the same time market these cattle at a younger age to
compete with cheaper western stock. This brought changes in
feeding practices, financing, and marketing. Men of tradi-
tion and experience shifted to intensive feeding of improved quality beef over a shorter period of time. The
dressed beef trade, which made it pay to produce quality,
helped Illinois and Iowa cattlemen who were willing to adjust
from being caught in the same position as Ohio feeders in
the 1850s who faced lower cost competition to the west. The
range cattle boom also created problems for the industry because of overexpansion and the decline of prices in the 1880s
and early 1890s, which caused some in the Middle West to turn
to other farm products.[1]
       With the end of the Civil War and the abnormal demand
for wool and the continued poor results with wheat in older
areas, many Illinois and Iowa farmers turned to growing feed
grains which they could sell or feed to their own stock. New
farms in northwestern Iowa, eastern Kansas and Nebraska, and
southern Minnesota added to the abundance of wheat and corn.
This contributed to a decline in grain prices, which gave a
relative advantage to stockmen who could utilize feed grains.
The agricultural press had been campaigning for some time

(and continued through the next decade) for farmers to "raise corn always in preference to wheat. Learn to convert Corn into Pork [and] Beef," farmers were told. Inclusion of livestock in the farm system became the watchword of agricultural writers such as "Tama Jim" Wilson and Coker F. Clarkson in Iowa and editors of the Prairie Farmer from Chicago.[2]

Civil War demand had, however, reduced the number of cattle in the Middle West. At the end of 1863 the Commissioner of Agriculture estimated the number to be down by 30% from the previous year in Illinois, the leading cattle-fattening state. In 1865 the Prairie Farmer assured midwestern farmers they had no rival for beef production in the East at least. They were urged to replenish war-depleted cattle resources in preparation for anticipated meat demands from the South and the growing population of the North. "Self-interest demands that strenuous efforts be made to increase [cattle] numbers."[3]

Cattlemen obtained their supply in a variety of ways. The most common practice was to raise cattle from calves on the farm, but most large-scale feeders had neither the capital nor the desire to maintain the number of brood cows necessary. Even so, an Iowa farmer contended that there was less personal satisfaction in buying feeder cattle than raising them.[4] Occasionally, a large-scale feeder raised his own cattle, especially before the 1880s. John D. Gillett, one of the most successful fat-cattle feeders in Illinois, raised a strain of Shorthorn grade cattle on his 12,000-acre property in Logan County. In 1876 Gillett had 500 cows and fed 700 fat cattle a year.[5] To supplement his own calves, Gillett bought cattle from neighboring farmers, one of the easier ways to obtain them without the expense of a large cow-calf herd. County reports in the newspapers frequently mentioned the scarcity of feeder cattle or that buyers had been active in the area taking surplus stock from farmers who did not specialize in feeding. Such local buying reached its height in the late summer and early fall when cattle came off pasture in good condition.[6]

The immediate neighborhood seldom supplied enough, so men went further afield into unsettled portions of the state where large herds grazed; into neighboring states; and to the Chicago, St. Louis, or Kansas City markets to buy feeder cattle shipped from adjoining states and the western range. Even before range cattle became available, Champaign County, Illinois, feeders bought poorer offerings in the Chicago markets and fed them up to a better grade. In 1876 Sangamon County feeders bought three-year-olds from Wisconsin, Iowa, and Missouri in the Chicago, Kansas City, and St. Louis markets.[7] In the 1870s much of northwestern Iowa remained open range and a source of feeder cattle. Farmers in the northeastern Iowa wheat area sold their surplus at one and two years of age to farmers in the southern half of the state and in Illinois, where corn abounded. The Iowa range supplied feeder cattle until the middle 1880s in competition with Texas, Indian territory, Kansas, Nebraska, and the increasingly important northern range—Wyoming, Montana, and Dakota. In the 1870s and 1880s midwestern farmers aided development of western ranches to which they sold stocker cattle and from which they later bought feeder cattle. Eugene S. Ellsworth (Iowa farmer-banker) brought cattle in the late 1880s and early 1890s from his Dakota ranch to farms in Emmet, Hancock, and Wright counties for fattening. As the business emerged from a period of depression in the 1890s and demand for feeder cattle increased, range men bought Canadian cattle as well.[8]

The source of feeder cattle and the place of market depended in certain areas upon the available railroads. Many Iowa men bought stock at the Missouri River markets where western railroads converged—Kansas City, Omaha, Council Bluffs—because it was cheaper than buying in Chicago and paying the freight back to Iowa. On the other hand central Illinois men found Chicago, where railroads running south and west converged, a better market for western feeder cattle. Absence of railroads limited feeding in some areas such as O'Brien County, Iowa, where no one fed western cattle until the railroad came through in 1881. In parts of Franklin County, Iowa, this was so until the advent of the Chicago Great Western Railroad in 1903.[9]

There was some argument over which was the most profitable method of operation, to feed fewer quality cattle or to handle a larger number of average steers. The Prairie Farmer and the Iowa Homestead agreed that the farmer who raised his own cattle, if of good quality, made the most money; but both newspapers assumed it was "better" in noneconomic ways for farmers to raise their own. As the Prairie Farmer put it, "the feeder has no business to buy his stock; he should breed and raise it himself."[10]

The movement of Texas cattle to northern markets after the Civil War was in response to the demand for more beef and the cheap price there. Though Texas cattle were not of major significance through the rest of the century as a supply, they were the first feeder cattle to come from outside the Middle West and were heralds of the later western range industry which came to supply the majority. As the cattle feeding industry became larger, the trend toward regional specialization continued; the range country raised cattle and Corn Belt farmers fattened them.

The business of shipping Texas and Cherokee (Oklahoma) cattle north resumed, rather than began, after the Civil War. The influx of Texas stock initiated range competition for Illinois and Iowa feeder cattle; and though the center of this competition moved to more northern ranges during the last quarter of the century, the consequences remained the same. Illinois and Iowa men adjusted to the presence of cheap range cattle by upgrading the quality of their beef and marketing it at a younger age.

Texas cattle reappeared in the North in 1866. They were shipped up the Mississippi River to Cairo and the Illinois Central Railroad yards or driven overland through Missouri. Supposedly, 10,000 Texas cattle were shipped to Chicago in 1866. George Duffield (a farmer from Keosauqua, Iowa) went to Texas in 1866 and bought 1,000 cattle at $12 a head and drove them back to Iowa during the summer. Less than half of Duffield's herd made it to Iowa in October because of stampedes and night losses, but the low price encouraged others to follow suit.[11] In 1867 Chicago received almost 40,000 head of Texas and Cherokee cattle, and the railroads and feeders prepared for a bigger year in 1868. Because of the low amount of shrinkage during shipment, a valuable hide, and lower initial cost, demand grew for southwestern cattle. The Illinois Central Railroad built a stockyard with a capacity of 4,000 head at Tolono, in Champaign County, where 15,000 Texas cattle were distributed to central Illinois feeders in June and July 1868. Estimates of the number of Texas cattle feeding in central Illinois that year went as high as 35,000.[12]

John T. Alexander, on his great Broadlands Farm in Champaign County, received shipments of Texas cattle numbering from 109 to 600 head every week from May 31 to July 26, 1868. Most came from Cairo, but some were from Abilene, Kans., via Chicago. Alexander paid $35 a head for over 4,000 Texas cattle the next year and after six to nine months on grass and some corn expected to sell them for $70 a head.[13] That he expected a profit of at least $100,000 indicates why many men wanted to process Texas cattle. They felt that a limited

amount of capital would buy more Texas than native cattle
for feeding; and though natives sold for more than Texans,
the profit per head was not much greater and the larger number of Texans produced more total profit. But by 1870 some
men, including Alexander, had second thoughts about the
profitability of Texas cattle. Morgan County feeders claimed
they lost money because of poor gains in weight, while Alexander apparently concluded that it was now a "losing business" and he would go back to feeding only native cattle.
Even so, by November 1870 Chicago receipts of Texans had
passed 200,000 head for the year; the northern fattened
Texas beef sold at a "good price," from $4.50 to $5.75 per
hundred pounds.14

Coincident with the reappearance of Texas cattle, the
southern or Texas cattle fever reappeared also and caused severe losses in native cattle as it had in the 1850s. Where-
ever natives mixed with Texans or occupied yards, roads, or
fields soon after, the native cattle usually died. Armed
Missouri farmers prevented Texas droves from passing through
to Illinois or Iowa and caused the cattlemen to seek better
ways to reach the northern feeding areas. In 1867 Joseph
McCoy (an Illinois cattleman) and others established a western railroad shipping point at Abilene, Kans. Texas drovers
could reach Abilene without crossing settled territory or
encountering many native herds, and the cattle interests
overcame local opposition by offering a bond for damages and
packing a public meeting in their favor. Though this procedure satisfied Missouri farmers, it did not prevent the
spread of Texas fever in Illinois from cattle brought by way
of the Mississippi River to Cairo.15

Native cattle in the same field with Texans died about
two months after contact. Cattlemen found it difficult to
understand the nature of the disease because it did not af-
fect the Texas cattle, yet there was an obvious connection
between their presence and its incidence. Consequently, it
was difficult to suggest remedies or propose procedures for
handling Texas cattle short of excluding them. The Illinois
legislature passed a quarantine law in 1867, but it proved
ineffective and native cattle losses increased in 1868. In
that year, Champaign County, Illinois, lost $150,000 worth;
Ford County lost 500 head, Grundy County 100 head. Vigilance
committees in some areas resorted to "calling on" Texas cattle importers to see that they made restitution for native
cattle lost.16

In 1868 amid charges, countercharges, and calls for

more information about the disease, a convention of cattle commissioners from twelve midwestern and eastern states and the province of Ontario met in Springfield, Ill. Eastern states had already (or were preparing to do so) quarantined live cattle shipments from Illinois in order to prevent the spread of the disease. Joseph McCoy visited the convention as an honorary delegate and did his best to defend Texas cattle and prevent exclusion as the only recommended means of combating the fever.[17]

The convention debates produced reports on the latest observations of the fever and prevailing opinions as to the cause and the treatment to be followed. There were three views on the subject. Texas and Kansas shippers refused to acknowledge the existence of a fever spread by Texas cattle. They admitted that northern native cattle died, but it was argued this was the result of a constitutional weakness. The "superior" Texas cattle, they pointed out, never died of the fever. Practical northern cattlemen assumed that ticks found on Texas cattle were related to the problem. Professional medical doctors and veterinarians, however, dismissed the cattle tick and sought the cause in spores and their transfer by breath or waste matter from one animal to the next. In 1868 a report submitted to the Illinois State Agricultural Society contained all the needed information about the fever and ticks to connect the two conclusively but concluded that the tick theory was the "last and perhaps least" of various explanations.

All reports and discussions at the convention made it clear that fever appeared only from contact with Texas cattle or places where they had been, that they carried ticks if shipped from Cairo on the railroad, but that such cattle coming from Abilene hardly ever had ticks because adults fell off during the long drive (which may account for the relatively little excitement over Texas fever in Iowa).[18] In central Illinois, at least, no Abilene cattle had brought fever to natives in the area. Discussion revealed the facts that ticks grew to large size on cattle, dropped off, and hatched young that infected pastures and climbed on cattle to suck blood and repeat the life cycle. John T. Alexander's farm manager had removed Texas cattle to a second pasture after the adults had fallen off and before young ticks climbed back. Native cattle in the first pasture subsequently died, while those in the second pasture contracted no fever from the Texas cattle without ticks. Native cattle pastured with tick-infested Texans for two weeks but removed before the ticks fell off did not die of fever, but those remaining in the lot four weeks or more subsequently contracted the disease. Men observed that frost in the fall arrested the fever and that any Texas cattle wintered over in Kansas did not carry it.[19]

When the medical doctors (not veterinarians) and cattle commissioners from various states presented their observations and opinions, the solid conclusions that could have been drawn about ticks were lost in a morass of contemporary medical ignorance, misinformation, hearsay evidence, and elaborate theories founded on only one observation. Frequently, under questioning by members of the convention, that one bit of "evidence" for a theory turned out to be inaccurate. The whole procedure illustrated the power of the "expert" in the form of medical doctors to perpetrate theories that might have been intriguing or in vogue but ran counter to the logic of the observed evidence. The most famous of the medical experts, the English veterinarian John Gamgee, headed a scientific commission to investigate Texas fever and his ably written and deceptively exhaustive report of 1871 set back research on Texas fever by a decade. Gamgee said that the tick theory "was not only impossible but ridiculous. A little thought should have satisfied anyone of the absurdity of the idea."[20]

In fairness to Gamgee, it should be mentioned that to advocate the tick theory with the prevailing knowledge of veterinary medicine, biology, and pathology in 1870 would have made him a laughingstock in the medical world, which did not yet understand the function of an alternate host. It was all the United States Bureau of Animal Industry could do in the 1890s to gain acceptance of the tick theory as fact. The first bulletin of the bureau published the findings of Theobald Smith and F. L. Kilbourne that a protozoan parasite in the blood caused Texas fever but that the parasite could travel from one animal to another only through an intermediate host, the tick. A study of the life cycle of the tick showed that it could be eliminated by dipping cattle periodically. By 1898 all restrictions had been removed on transportation of Texas cattle if they were dipped.[21]

But the solution came long after the immediate problems of 1867 and 1868 when the only seemingly feasible response was a quarantine on Texas cattle from spring until the fall frosts. The convention of cattle commissioners so recommended, much to the disgust of McCoy, who saw it all as a plot to ruin him and the dealers in Texas cattle and to protect native cattle raisers. McCoy labeled the convention, perhaps rightly for the wrong reasons, "as a collection of quondom quacks and impractical theorists and imbecile ignoramuses, . . . without an equal." The medical men especially brought to McCoy's mind "the ancient royal commission of sage scientists who spent many days and weeks investigating and profoundly debating the all-absorbing question of natural history, to wit: 'which is the butt end of a billy goat?'"[22]

McCoy was not alone in his feeling that the law was being used to the advantage of native cattle feeders, but pressure favored the local interests; and in 1869 the Illinois legislature passed a bill excluding Texas and Cherokee cattle from March 1 to November 1, with the exception of cattle wintered in the states of Kansas, Nebraska, Missouri, Iowa, or Wisconsin, attested by a certificate from the clerk of the county where they were wintered. As a result, Kansas county clerks issued certificates liberally.[23] The debate over Texas fever continued to the 1890s, but the outbreaks became less severe after cattle coming from Texas to Cairo by boat were more effectively prohibited by the 1869 law or segregated, if allowed through, under later procedures. Texas cattle that came by way of Kansas lost most of their ticks even if fraudulently certified and so posed less of a danger to native cattle. Feeders also made efforts to segregate Texas cattle, though outbreaks of fever occurred in Illinois in McLean, Morgan, and Sangamon counties in 1872; Champaign County in 1876; and Perry County in 1885 and in the 1890s.[24] Railroads disregarded the law, which was eventually held unconstitutional by the Illinois Supreme Court in 1879 after a ruling by the U.S. Supreme Court that a similar Missouri quarantine law was an unconstitutional restraint of interstate commerce. Some cattlemen thought they could make a lot of money buying Texas cattle cheaply and selling them in Chicago after a summer on Illinois and Iowa bluegrass, and they did not allow quarantine laws and subsequent proclamations of state governors to effectively hinder their importation of Texas cattle to the Middle West.[25]

By 1886 Texas cattle could be brought into Illinois legally for immediate slaughter, if segregated from other cattle while being shipped, and for feeding, with permission of the state livestock board after a 90-day quarantine.[26] Also by the 1880s range cattle from Kansas, Nebraska, Wyoming,

Dakota, and Montana replaced those shipped directly from Texas and Cherokee country as feeder stock.

Texas fever killed many native cattle and disorganized the cattle feeding business in particular areas. Less dramatic but more significant, the competition of Texas and western range cattle brought changes in Illinois and Iowa beef feeding practice. This was a matter of choice to the feeder depending upon how much corn or grass, capital, and native cattle he had. Beef canners and packers bought western cattle as a cheaper more abundant source of meat than the better grades of natives used for live exports and dressed beef. There was a fairly good demand for western feeder stock, and even western grass-fattened cattle were sold directly to packers without an intermediate stop in the Middle West. The growing size of this latter group, which went to market from June through November, caused problems for Corn Belt feeders. In 1867, for example, 120,630 Texas and Colorado cattle arrived in Chicago during the six-month range season. This amounted to about 10.9% of total cattle receipts for that year.[27] From 1882 to 1895 Texas and western receipts in Chicago fluctuated between 36% and 25% of the total. The number dropped off sharply after 1895. In 1900 range cattle accounted for only 13% of total receipts. (See Table B.12.) Generally in the 1880s and 1890s, prices for western cattle ranged from about 16% to 25% less per hundred pounds than for native 1,200- to 1,500-pound steers. Prices of Texas cattle were the same as western cattle or a bit lower. Texans wintered in the north sold for more than direct shipments from the south. Northern-wintered Texans also created some problems in nomenclature in the market, Were they Texans or westerns?

As early as 1868 men began to discuss the consequences of western competition. The cheap cattle weakened prices, especially for the poor offerings of native cattle, and editors and market reporters urged feeders to hold their stock until properly fattened at the age of four years or more when they would bring better prices.[28] But the remedy lay not with fattening common stock at a later age, but in upgrading the quality of midwestern cattle and fattening them at a younger age for a faster turnover of capital. In 1870 the Department of Agriculture anticipated that within a decade there would be "an immense improvement" in midwestern stock as the "more intelligent cattle-raisers" supplanted their common native cattle with improved breeds. Individual complaints supported the department's view that native un-improved stock would become unprofitable for Illinois and Iowa feeders.[29]

As Illinois and Iowa land values increased, men who fattened cattle on grass were hard pressed. Many felt they must turn to grain to realize a sufficient return per acre. As early as 1872 some Morgan County, Illinois, grazers broke up their pastures and planted corn.[30] The Commissioner of Agriculture suggested in 1876 that the best way to organize the resources of the Middle West and the range country was to raise cattle to two years on the ranges and then send them to the Middle West for a year's fattening on corn.[31] This is what Illinois and Iowa feeders were doing in the 1870s and early 1880s. They fed improved local and range cattle and sent them to market at three years instead of four, while those who did not improve their stock turned to grain farming. By 1880 the Illinois Department of Agriculture reported that only the "choicest specimens of early matured steers" would successfully compete with the better grade range cattle.[32] As cattlemen improved their range herds with Shorthorn and Hereford bulls and shipped increasing numbers east, the competition, together with the developing centralization of the packing industry, lowered average prices in Chicago from an 1882 high of $6.25 per hundred pounds (1,200- to 1,500-pound steers) to $5.90 by 1884, $4.70 in 1888, and the low of $3.90 in 1889. Prices did not go above $5 per hundred pounds again until 1900.[33]

The agricultural press and farm leaders continued to exhort beef feeders to improve their animals and feeding procedures; they assured that "scientific and methodical application" of principles of good feeding "must necessarily serve to a large extent in overcoming the disadvantage of raising stock in high priced lands."[34] Writers stressed the ultimate advantage of combining Illinois and Iowa corn and improved cattle "as there will always be a demand for them at high prices"; and urged that only "good cattle" fed for the top market would do for the corn states.[35] As profit margins narrowed, it became more important to have improved cattle that would gain 100 pounds on fewer bushels of corn than the average available in the past. The profit came from cattle that converted corn into beef most economically.

Financial arrangements of feeders were an important aspect of the business. Increasing specialization in the production of prime beef cattle brought changes in the financ-

ing of feeding operations. Considerably more capital was involved in feeding improved grade cattle than in handling natives or Texans. Toward the end of the century the need for
extra capital helped the development of cattle loan companies and the business of city banks discounting cattle loans
made by rural banks.

In the years just before and after the Civil War some
packers advanced money to feeders. Isaac Funk had occasionally fed cattle under contract to a packer who supplied the
capital. But this could not have accounted for much of the
total number fed because packers experienced chronic short-
ages of operating capital as they expanded their operations
after 1875. Alexander, Strawn, and others before the 1860s
borrowed extensively from local farmers or took cattle on
credit and paid for them after they were marketed.[36] In the
last half of the century much of the operation might have
been similar to this. In some areas men who wanted to feed
more cattle than they could raise or buy with their own
funds could borrow from well-to-do farmers if their reputations were good.[37] Unfortunately, this is the least documented
side of cattle feeder financing.

Country bankers backed some of the sizable operations.
George Wilson, a Geneseo, Illinois, banker "with that rare
quality called 'cattle sense,'" backed Alexander's purchase
of 3,000 cattle during the panic of 1873. Local buyers supplied feeders on credit to trusted individuals, which allowed farmers without capital to feed cattle.[38] Rural sources
of capital seemed adequate until the late 1880s when cattle
loan companies appeared, many of which were backed by the
packers or stockyard banks and commission firms. The financial panic of 1893 and its aftermath placed an added strain
on rural credit. Men found it difficult to borrow from normal local sources. In 1896 a Prairie Farmer inquiry found
that local banks were short of funds and more men would be
feeding cattle if they could borrow money.[39] A more formal
arrangement came with the new loan companies. Feeders had to
secure loans with chattel mortgages on the cattle rather
than a personal note. They also had to supply financial information about themselves and sometimes have their operations inspected before their personal reputations were established with the loan company. By the early twentieth century
larger loans come from cattle loan companies while small
feeders borrowed from local banks that sold the cattle notes
to a city bank. Even with the more formal arrangements in-

dividual reputation played a great part in securing loans.40

The railroad was the major means of moving Illinois and Iowa cattle to market after 1865, and although they reached market faster than on foot, the problems of loss of weight and damage from overcrowding still existed. The first cattle cars were a makeshift lot that did not provide adequate protection. The air brake and an improved coupler helped conditions in the 1870s, but overcrowding and crippling continued. Concern over excessive loss of weight (as high as 15% of gross weight on the run from Chicago to New York) caused the enactment of the federal 28-hour law in 1873. This law limited interstate shipment of animals to a period no longer than 28 hours without unloading for food, water, and rest for a period of five consecutive hours. The law was defective and remained a dead letter until about 1905.41

The fluctuating rail rates between Chicago and New York affected prices paid to Illinois and Iowa feeders by livestock shippers or the dressed beef packers. Buyers effectively passed the higher rates on to the feeders by bidding less on cattle to be sent east. Periodic relief came in the form of rate wars between competing rail systems, but little relief was likely for the Illinois or Iowa farmer who had no choice but to ship on the railroad that served his area. The most frequent complaint against railroads, however, concerned the shortage of cars or the length of time for the trip to Chicago. Though the accusation was denied, feeders claimed that the railroads perpetuated the three-day market in Chicago by not furnishing cattle cars on an even distribution through the week. The railroads made efforts to come to agreement among themselves as to time of runs and rates from western points into Chicago. If they did not agree on rates, they often agreed on lengthening the time of the run, which in effect increased the rates.42

The three-day Chicago market was another problem for feeders. Most receipts came into Chicago on Monday, Wednesday, and Thursday with Monday being the dominant day. The uneven distribution of receipts strained market facilities, put uneconomical pressure on railroad equipment, and gave buyers a price advantage. With the bulk of weekly receipts coming on Monday and Wednesday, buyers could beat down prices by waiting until the yards filled up and men became worried about carrying cattle over to the next day. Feeders said that the railroads would not supply cars or put cattle on through trains except for the Monday or Wednesday markets. On the other hand, some feeders claimed that they received better prices on days of heavy receipts because of competition among buyers and suffered financially every time they tried to even market receipts. The railroads and the packers insisted they preferred a five-day market but that the farmers refused to even shipments if given the opportunity. There the matter stood until the government forced changes by a market zone system as part of World War I food controls.[43]

Farmers could lessen their costs by shipping carload lots and obtaining the most recent market information. Very little accurate information was available to the end of the century about how many cattle were in feedlots across the Middle West. Feeders and packers alike operated without sufficient information to plan very far ahead. Newspapers occasionally conducted surveys to estimate the number of cattle on feed and then advised their readers.[44] The Illinois Department of Agriculture in its crop report bulletins included assessors' returns of cattle on hand on May 1 as well as a list of the estimated number of cattle marketed and on hand in December compared to the previous year. This was of little value, however, because the estimate was always placed at 20% of the number on hand regardless of other factors. It was far better to do as William Stevenson did. This Tama County, Iowa, feeder subscribed to the weekly Drovers' Journal and received the latest market reports by telegraph the morning he intended to ship cattle to Chicago. He did business through the commission firm operated by the brother of "Tama Jim" Wilson (well-known Iowa farmer and U.S. Secretary of Agriculture).[45]

A good commission firm was probably the feeders' best source of market information despite attacks by the Grange and newspapers. In the 1860s when some commission firms sold cattle for country buyers, these firms tried to prevent reporting of stockyard sales prices in order to keep farmers from having a source of information other than the local buyer. The Prairie Farmer retaliated by printing the names of members of the "Ring" to keep sales secret so that farmers would know who not to patronize.[46] Commission firms came in for much abuse as a monopoly and were frequently blamed for problems beyond their control. In the mind of the militant agricultural press, the commission houses were part of the Union Stock Yards monopoly, which was on the opposite (and wrong) side of every issue from the standpoint of the

67

meat producers. Through most of the last half of the century, the commission firms charged 50 cents a head for handling and selling cattle. This did not seem an exorbitant charge on a steer worth about $55 in the late 1890s.[47]

The better cattle feeders in the Corn Belt tried to produce prime, choice, or good steers, while the majority produced good or medium steers and indifferent feeders sold common grades. Cattle grades set at the stockyards were still not uniform and precise by the end of the century;[48] but there seemed to be general agreement among buyers for the packers and livestock reporters on the major characteristics of each grade. The grade assigned in the stockyards was important because it affected the price. Men had to be able to judge which grade their feeder cattle could be prepared for. Poor quality animals could not attain a prime grade no matter how they were fed, and the cattleman who could not see the limits and possibilities of those he bought would soon lose money. It cost more to feed cattle for the top two grades than for good or medium grades. There were usually too few of the top two grades and too many of the bottom three grades of beef on the market. The desirable feeder cattle for top grade beef had short legs, broad flat backs, and straight top and under lines that were nearly parallel. When the animal was properly fed, the outline remained the same while weight and total size increased. Lesser grades departed from the desired image in greater degree as one descended through the list, with lower grade cattle having longer legs, thinner hips and chest, and higher flanks. By the end of the century, few Illinois and Iowa cattle in the better grades exceeded 1,600 pounds, and many prime baby beeves went to market at weights from 800 to 1,100 pounds. This contrasted with the 1850s and 1870s when extra-fine cattle were four- to six-year-old steers and heavyweights over 1,500 pounds received better prices.[49] Western competition had forced Illinois and Iowa feeders to concentrate on quality beef cattle because range men could feed common to good steers on grass at less cost than midwestern farmers could.

Not all feeders sold fat stock of the quality necessary to compete successfully with range cattle. Prairie Farmer market reports frequently noted the lack of first-rate beef in the market and the oversupply of half-fattened stock; the editors urged feeders to ship only fully fattened cattle. In periods of drought or poor corn crops many feeders did not want to carry stock over the winter and therefore sent it to market early and in less than prime condition. Astute feeders profited from poor planning on the part of others and bought these partially fattened cattle and fed them a short time for the midwinter market.[50]

Illinois and Iowa men who fattened cattle on grass sold them in the fall in competition with range cattle and in the late spring before the latter came onto the market. In the fall they needed choice or prime quality animals to compete successfully.[51] Corn-fed cattle could go to market at any time they were ready, though this was generally during the winter and spring.

During the 1880s and into the 1890s corn feeding increased over grass, and the ages and weights of marketed fat cattle gradually declined. That is, the prime steers were more in the two- and three-year-old range and generally less than 1,600 pounds, while for the common and medium grades average weights probably increased from the previously low level.[52]

Although accounting for only a small percentage of the market, prime and choice cattle two years old or less, known as "baby beef," came to be a specialty of leading Illinois and Iowa cattlemen. The result of years of breeding and feeding for early maturity, these were prime examples of what corn and feeding skill could do to compete with range cattle in a market of falling prices. The Iowa Homestead said in 1891 that "the time will come, and it may not be very far in the future, when it may be desirable to feed everything off at from eighteen to twenty months." Leading cattle feeders produced baby beef by starting with top-quality calves and feeding them as much grain, grass, and supplements as the animals would take. In December 1895 ten 1,200-pound steers twenty months old brought the top price in the market. The next year heavy steers over 1,500 pounds sold at a discount regardless of grade.[53]

The majority of Illinois and Iowa cattle feeders did not produce baby beef nor always send prime or choice cattle to market. However, they were influenced by the general trends that more often than not made grass feeding and four-year-old cattle unprofitable in the Corn Belt. Joel W. Hopkins (of Putnam County, Illinois) was an example of prosperous Corn Belt farmers who fed cattle as a major part of their farm operations. By the end of the century, cattle feeding was not seasonal. Hopkins bought and sold cattle in practically every month of the years 1897 through 1899.

69

In 1897 Hopkins bought 797 cattle and sold 702. The next year, he bought 482 cattle and sold 515, and in 1899 he bought 607 and sold 641. A good many were bought and sold locally, but Hopkins periodically sold one and two carload lots in Chicago (carload = 16 cattle). In 1897, he sold one carload each in February, April, and May; two carloads in March, June, November, and December; and six carloads in July. He generally received from $4.45 to $4.65 per hundred pounds, but one March carload sold for $3.75 and the December lot sold at $4.80. At no time did prices of his steers exceed the reported high price for the month, but they were about average. The other two years were not notably different except that, with the general trend, prices received moved higher into the $5 and $6 range. Where shown, average weights of Hopkins's steers ran from 1,300 to 1,400 pounds. Hopkins bought about as many cattle in Chicago as he sold there, but he also bought and sold lots as large as 20 head locally.[54]

Hopkins had been feeding cattle for years and had not curtailed his operations as some did during the decline in cattle prices after 1882. Average beef cattle prices had risen in 1882 to the high of the last quarter of the century because of the growing demand for meat as well as the demand for cattle to stock ranges. But the trend soon reversed as the range filled and began sending large numbers to market. Overexpansion and periodic liquidations in the range cattle business and increasing consolidation of the packing industry kept the price trend moving downward after 1882, and the business depression pushed prices lower in the early 1890s. Not until 1896 did the trend of beef prices start up again; in 1902 they surpassed the 1882 high. (See Fig. B.2.) Within the long-term price trend, there were seasonal variations, with declines in the summer and fall at the time of heaviest receipts from the range and rises from December through May or June. The increase in corn fattening of range cattle in the 1890s diverted some from market in the fall and added to market receipts in the spring. This tended to reduce seasonal variations in prices.[55] The price rise in the 1890s came after a long period of low prices and a two-year liquidation of feeder stock in 1893 and 1894 when feed had been scarce. From year to year, prices could be affected by the size of the corn crop. Cheap corn encouraged men to feed it to cattle and swine; however, a high price encouraged less demand for feeder cattle or for the marketing of fat cattle at lighter weights.[56] The extra-large corn crop of 1895 increased the demand for available feeder cattle, which pushed prices almost as high as those for fat cattle. This further depletion of feeder stock supplies meant a shorter supply of fat cattle, and this brought market receipts of them more in line with demand. By 1896 cattle prices started a steady climb upward. Chicago receipts had reached a fifty-three-year high of 3,571,796 in 1892 and then declined and did not surpass that figure until 1918.[57]

In the 1880s and 1890s some Illinois and Iowa cattlemen claimed that they lost money. In 1886 W. Whitson (Pottawattamie County, Iowa, feeder) said he lost $600 feeding cattle in the previous year and thought he would lose money again. In 1898 E. E. Chester (at the Champaign County, Illinois, Farmers' Institute) claimed that cattle feeders in the county had lost money for the previous five years. Many of the men in east-central Illinois who had fattened cattle on bluegrass gave it up and turned their pastures into cornfields in the late 1880s and early 1890s. On the other hand, James N. Brown's sons said that they only fed cattle because they had bluegrass pastures, but that it would not pay them to fatten cattle on corn. The Bureau of Animal Industry concluded in 1885 that livestock feeders generally gave "little careful thought" to figuring operating costs.[58] It was also evident that individual ability made the difference between profit and loss during periods of low prices. Prime cattle still brought good prices for those who could produce them. J. W. Steele (Monona County, Iowa) made $16 a head on a carload of steers sent to Chicago in 1888. Fred Stowe (Grimes, Iowa) sold five carloads of cattle in 1891 at fully $2 per hundred pounds above average prices for the month, and others did the same. In times of good prices the rewards for prime cattle were even greater. In 1899 L. H. Kerrick (McLean County, Illinois, and son-in-law of Isaac Funk) sold two carloads of grade Angus steers for $8.25 per hundred pounds. These prime cattle brought 75 cents more than the highest price paid in 16 years and almost $2 above average prices for the month.[59] Good cattle feeders could get better than average prices in almost any market.

Many suggested that the Chicago packers caused the price decline. The Iowa Homestead claimed that the "dressed beef combine," when possible, forced down the price of good beef to the level of range cattle prices.[60] A U.S. Senate committee investigated beef prices from 1888 to 1890 and

71

charged that if men refused the first bid on their cattle in the Chicago market, the packer buyers would later force them to accept a lower one or offer none at all. The committee found that whether by design or not, packers acted together in buying cattle, setting the retail cost of beef, agreeing on retail market areas, bidding on government contracts, and underselling local butchers. The committee disputed Philip Armour's opinion that low prices came from too many cattle in the market by pointing out that in the fall of 1886 when receipts fell, prices fell, and that in 1890 low beef prices existed at the same time that there were fewer beef cattle in the country, a larger population, and higher exports than in the early 1880s.[61]

The Union Stock Yard's Annual Report and the market columns of the Prairie Farmer followed the more orthodox version of the price decline; that is, that it came from over-expansion on the range, large receipts of medium to poor grade cattle in the late 1880s, and the generally poor business conditions of the early 1890s. With the decline in the number of cattle and the increasing consumption of beef as economic conditions improved, prices began to rise after 1896.[62] In 1897 A. B. Groat (president of Illinois Cattle Breeders' Association) pointed out that the "good old days" of cattle feeding were gone and that farmers had to make the most of the fact that they had no effective control over prices. Feeders, therefore, must pay more attention to the costs of production, quality of cattle, and management of the feeding program.[63]

Cost-conscious feeders must assess the situation as to the price of feeder cattle, number on feed, cost of feed, and probable level of fat-cattle prices at the time they expected to market their animals. About the only figure they could ascertain with certainty was the current cost of feeder cattle. Feed costs over the months varied with size and quality of corn and fodder crops; other variables they could only estimate. A dry summer on the range might flood the fall market with under average quality cattle and bring down prices. A mild or severe winter could make a difference of about two months either way as to when fat range cattle first came to market in the late spring and early summer and would have an effect on prices. Feeders could have more than one carload of cattle on feed so that they could pick and choose from the feedlot to make up a load of fairly well-matched cattle. A carload of even quality beef brought a better average price than one of mixed quality. A few poor quality cattle in a lot would bring down the lot price proportionately more than a few above average cattle would raise it. Also the feeder could sell his carloads directly through a commission merchant in Chicago rather than selling a few animals at a time to a local buyer who paid less than the Chicago price in order to make his own profit.

By the end of the century Illinois and Iowa cattle feeding had changed from beginning the fattening of cattle for market after the age of three to completion of the process almost always before that age. A small but growing percentage of fat cattle were going to market at 18 to 24 months by 1900. Most of these young animals had a high percentage of purebred blood as did most of the feeder cattle raised in Illinois and Iowa. The range cattle brought into the feedlots and fattened on corn had a higher proportion of improved blood than those of two decades earlier and were far superior to the range cattle from Texas and Oklahoma that appeared in the 1850s and 1860s.

By 1900 several methods of cattle feeding operations were followed in Illinois and Iowa. Some men bred and raised their own high-grade cattle for fattening. Another large portion of feeders bought young range cattle in the Kansas City or Chicago markets and fattened them, and a third group fattened cattle raised on the farm as well as those brought from the range. But whether from the range or calved on farms, the majority of top-grade beef animals spent some time in feedlots eating corn.

Range competition and declining profit margins in the 1880s and early 1890s caused reactions among Illinois and Iowa feeders and caused many of the Illinois men who had fattened cattle on bluegrass to plow it up and turn their attention to grain farming. Nevertheless, efficient cattlemen in Illinois and Iowa combined their cheap corn and knowledge of fattening with the young western range cattle to continue producing beef at a profit. These men had survived by the end of the century because they produced cattle to match the changing market structure.

CHAPTER 7. REGIONAL CONCENTRATION
OF ILLINOIS-IOWA CATTLE FEEDERS

In the last half of the nineteenth century changes occurred in the areal concentration of cattle feeding operations, which shifted from central Illinois to northwestern
Illinois and central and southwestern Iowa. Though only a
partial measure, federal census statistics give some indication of the distribution of beef cattle on a county basis
and reveal changes in this density in Illinois and Iowa.

From 1850 to 1890 the decennial census takers counted
oxen, dairy cows, and "other" (beef) cattle on farms as of
June 1. This generally excluded spring calves, but not always, depending upon how local enumerators interpreted
their instructions. In 1900 the census schedules further
subdivided cattle into calves under one year of age, steers
from ages of one to two, two to three, and three and over,
bulls one year old and over, heifers from one to two years
old, dairy cows over two years old, and other cows. Because
of the new divisions, figures for 1900 are not fully comparable with the earlier years. Also the June 1 date posed
problems as a guide to feeder stock on farms because many
fat cattle had already gone to spring market. In addition,
Illinois tax assessors came in May or June, which induced
pennywise farmers to have fewer cattle at that time of year.
Within the limitations of ten-year intervals, a June 1 reporting date, and the county as the reporting unit, census
figures for beef cattle per one hundred acres gave some indication of the magnitude of cattle feeding across Illinois
and Iowa.[1]

In 1850 the highest density was in west-central Illinois
in Morgan County, which had 4.7 beef cattle per hundred
acres. (See Map A.7.) Scott County, immediately to the west,
was second with 4.2, while surrounding counties to the south,
east, and north had over 3. The eastern range counties of

74

Coles, Edgar, and Vermilion and the northeastern counties had from 1.5 to 3.1 beef cattle per hundred acres. McLean County in central Illinois, already noted for its corn and grass and settled by such cattlemen as the Funk family, had 2 beef cattle per hundred acres. The area from Morgan County south to Randolph showed a concentration from 1.5 to 2.7, while the area to the west and north of Morgan had more than 1.5 with the exceptions of Hancock, Mason, and Schuyler counties. Four counties in the southeast corner of Iowa had more than 1.5 beef cattle per hundred acres. These Iowa counties were the major area of older agricultural settlement.

In 1860 counties did not show major increases over the highest number of beef cattle in 1850, but many new counties had more than 1.5 per hundred acres. (See Map A.8.) The highest density was in Kane County in northeastern Illinois, which had 4.9 per hundred acres, while three southeastern Iowa counties had passed the 3.1 mark. Southern and northwestern Illinois counties showed increases in beef cattle, in many cases doubling the number since 1850. Increased settlement and the advance of the railroads accounted for much of this growth. In 1855 Benjamin F. Harris estimated that the construction of the railroad advanced the price of livestock 10% in one year.[2]

The Iowa growth was more uniform in progressing outward from the southeast toward the west as far as Appanoose County and north as far as Dubuque County to form an arc (turning on Washington County) with a density of from 1.5 to 3.1 beef cattle per hundred acres.

In 1870 Illinois had begun to consolidate into the general pattern it would follow for the rest of the century. (See Map A.9.) Morgan County dominated, but in the second magnitude the balance had shifted to north and west of the Illinois River. The number of beef animals in southern Illinois declined, and from then on that portion of the state never figured prominently in the fat-cattle business. McLean, Logan, and Sangamon counties in the center of the state still stood out from the surrounding area as centers of activity. In eastern Illinois, there was another center of concentration on the fringes of the upper Wabash valley. Iowa continued much as before as it shifted the orders of magnitude one step to the west. The old arc encompassing counties with from 1.56 to 3.10 beef cattle per hundred acres now generally contained from 3.12 to 4.66. Three counties (Clinton,

Louisa, and Muscatine) passed the 4.6 mark. Four western
Iowa counties in the Missouri River valley formed an area of
beef cattle concentration not contiguous with any other with
the same number of "other" cattle. (See Table B.13.)

In 1880 northwestern Illinois had become the dominant
part of that state, though McLean, Morgan, and Sangamon
counties stood out in the center and the prominence of the
upper Wabash valley area was still noticeable. (See Map
A.10.) Carroll County in the upper northwest took the lead
in Illinois with over 8 "other" cattle per hundred acres.
Iowa no longer fit the arc description because the east-central counties of Cedar, Clinton, Delaware, Jones, Linn, and
Muscatine formed the dominant area. Marshall County in the
center of the state had the same number. The southwest counties again showed a higher concentration of cattle than the
surrounding region.

From the Illinois map for 1890 a continuation of the
trend toward the northwest can be seen. (See Map A.11.) The
central and eastern counties increased in absolute numbers
of beef cattle, but so did the northwestern counties and in
eight cases by two steps in the scale. In the west-central
part of the state Cass, Christian, Menard, Morgan, and Scott
counties increased by more than one step, while in the east-
central part, only De Witt County increased by two steps. By
1890 in some southwestern Iowa counties, the number of beef
cattle doubled and even tripled over the number in 1880.
While Illinois had only two counties with over 10.9 beef
cattle per hundred acres, Iowa had 26 counties with over
10.9. The area of highest concentrations in Iowa stretched
from Clinton County in the east to Jasper County in the center and then to the southwest through Warren, Clarke, and
Ringgold counties to the Missouri River. The density patterns graphically show what census statistics reveal; Iowa
had become the leader of the two states in numbers of beef
cattle. This represented a continuation of the pattern noted
earlier: the area of cattle feeding shifted west as eastern
Illinois farmers concentrated on more cash-grain production.

By 1900 the number of beef cattle had declined in nine
counties of central and eastern Illinois and in the west-
central Illinois counties of Morgan, Sangamon, and Scott.
(See Map A.12.) In several of the northwestern counties
numbers had increased, but none had over 12.5 per hundred
acres. In Iowa, the number continued to increase in many
parts of the state. Twenty-seven counties had over 12.5 beef

cattle per hundred acres and eleven of those had over 14.
Though there were two areas of concentration in the eastern
part of the state, southwestern Iowa had become the major
center. The concentration also extended to the north as far
as Cherokee, Buena Vista, and Pocahontas counties in the
western half of the state and to Franklin County in the eastern half. The northern counties of Osceola, Lyon, Kossuth,
Palo Alto, and Allamakee contained the fewest beef cattle
per hundred acres as they usually had in each of the previous decades.[3]

Some of the reasons behind the areal distributions of
beef cattle are relatively easy to find. In southern Illinois, the soil with a claypan underlayer did not drain well
and was considerably less productive than the prairie soils
to the north of the Shelbyville moraine. Farmers in the area
grew less corn and more wheat, kept fewer animals of poorer
quality, and made less money to invest in improvement. Few
breeders of improved cattle lived in the area and local
farmers thought such stock cost too much. As late as 1878 in
Wabash County some people ridiculed purchasers of improved
stock. P. E. Michaels (Jackson County farmer) wrote in 1880
that "I never was in a section of the country where the
farmers, as a rule, took so little care of cattle as they do
in this. It is no wonder that the cattle here are called
'scrubs,' for most of them are treated badly enough to make
scrubs out of any thing."[4] In general, the area south of a
line from Clark County in the east to Jersey County in the
west was less prosperous and less advanced in agricultural
technique than the rest of Illinois.

Mason County in west-central Illinois never had over
2.6 beef cattle per hundred acres because practically the
whole county was covered with a sandy soil of low productivity, and offered little attraction to settlers when there
was so much good land all around. Cook County in the northeast contained Chicago; its urban development hindered farming, though distilleries in the area fattened cattle on
mash.

East-central Illinois grazers became less concerned
with fattening cattle after 1880 as land values increased.
Owners profited by draining the land and planting grain. In
Champaign County, for example, from 1879 to 1899 organized
drainage districts encompassed 30% of the land area. Further
encouragement for the change to grain came when cattle prices
started to fall after 1882. The several railroads running

south from Chicago provided ready access to the city market, and grain elevators appeared at convenient intervals.[5] Owners of large tracts (especially in the second generation) who did not want to supervise large-scale livestock operations probably found it easier to rent their land to tenants. The tenants in turn found it easier to meet the rent requirements with a cash-grain crop than with livestock, and the landlords found it less risky to ask for a cash payment or shares of a grain crop.[6] Indeed, Thomas Scully (the great midwestern landowner) rented his Logan County lands for only one year at a time from January 1. This in addition to a cash rent requirement discouraged cattle feeding by his tenants because January 1 was not as logical a termination date for feeding operations as for grain production. Perhaps one reason Logan County did not have more beef cattle despite the example of John D. Gillett was that by 1890 52% (1900, 60%) of the farms in the county were tenant operated. As early as 1870 Scully alone owned 30,000 acres in Logan County, or about 7.56% of the land.[7] Successful cattle feeding required good judgment and skill in addition to steady work. While practically anyone could raise a corn crop in east-central Illinois, many could not fatten cattle at a profit in the 1880s and 1890s.

Not all landlords encouraged grain raising exclusively. Samuel Allerton (Chicago meat packer, businessman, and large-scale landowner in Piatt, Vermilion, and Henry counties, Illinois) encouraged his tenants to raise cattle in order to have a good rotation system between pastures and corn and to produce manure to help maintain soil fertility.[8]

The early Illinois concentrations of beef cattle occurred in Cass, McLean, Morgan, Sangamon, and Scott counties where from the 1830s men had been interested in fat and pure-bred cattle. James N. Brown and his sons in Sangamon County advocated bluegrass pasture and purebred cattle to all who would listen. The Browns, through sales of purebred cattle, provided a way to improve central Illinois stock as well as a visible example for cattlemen to follow.[9] Alexander came to Illinois from Ohio where he had learned the cattle business. He bought land in Morgan County in 1848 and almost immediately made a name for himself as a cattleman. In company with Strawn, Alexander made Morgan County well known in the New York and Chicago markets. During 1870 at the high point in his career, Alexander supposedly marketed over 50,000 head of cattle.[10] Strawn also learned the business in Ohio before he moved to Morgan County in 1831. Strawn was said to dominate the St. Louis cattle market for many years before 1850 and also sent cattle to Chicago, New York, and New Orleans. In the 1850s he was one of the first in the area to switch from grazing steers on prairie to feeding them on shocked corn. When he died in 1865, his estate was valued at about $1 million.[11]

Besides such well-known men as Alexander, Funk, Gillett, Harris, and Strawn, many smaller scale operators, to a certain extent influenced by these leaders, were the real backbone of the cattle feeding business. In Morgan County, for example, long-time resident John Henry credited Strawn's example with encouraging other farmers to feed cattle or to change their feeding practices. Strawn's "mode of farming was so simple that all could adopt it and see the great advantage derived from it."[12] Many in the Morgan-Sangamon county area came from Kentucky where there was an early cattle feeding center. The stall feeding method was often called the "Kentucky" method. For example, Samuel Wood came to Morgan County from Kentucky with his parents in the 1830s and became a cattle feeder and dealer, sometimes grazing 1,000 head of cattle.[13]

Other men fed cattle because their fathers had. John Burch was born in Morgan County in 1842 and reared on a farm where cattle were fattened. More representative perhaps than the better known owners of thousands of acres, Burch was a successful farmer and feeder who had accumulated 750 acres by the end of the century.[14] Immigrants unfamiliar with either Kentucky or Illinois cattle raising could follow the example of successful feeders. William Cleary, born in Ireland in 1818, came to Morgan County in 1838, where he worked as a farmhand and gradually bought land. Cleary sold his various bits of land in 1859 to buy one contiguous farm of a little over 500 acres and began raising stock and feeding cattle.[15] But even in counties with large numbers of beef cattle, not all farmers were feeders. Of the 557 men listed in an 1878 directory of Morgan County township (13 North, Range 8 West) 316 were identified as follows: farmer (187), farmer and stock raiser (57), renter (45), farmhand (26), and stock dealer (1).[16] The fact that stock raisers were mentioned separately indicates that they were considered different from farmers who raised just grain and perhaps a few livestock.

By 1880 the center of cattle feeding in Illinois had

shifted to the counties north and west of the Illinois
River. Some credited this to the fact that English farmers
settled parts of those counties and had "a fancy to improve
their cattle," while the Scots turned to sheep and Yankees
would do anything that paid. More recent studies question
the influence of cultural and national origins on type of
agricultural pursuits.[17] More influential in choice of type
of farming in both states, it would seem, were the examples
of leading farmers in the area who turned to the crop-livestock association that utilized their resources to best advantage and satisfied market demands. The more rolling terrain to the north and west of the Illinois River provided a
good combination of fertile cropland for corn and more slop-
ing areas for pasture. Illinois wheat met increasing competition from regions west of the Mississippi by 1870, which
increased the trend toward corn as a grain that could be
sold or fed.[18] It was possible too that those who grazed
cattle at first because of terrain and distance to grain
markets were so accustomed to handling them that they
switched to feeding more grain and less grass over time even
though their county was no longer primarily range. Large
corn crops and low prices also could have had a temporary
effect on the number of cattle on feed.[19] Mercer and Warren
counties were the leaders in cattle feeding according to the
census figures at the end of the century. In Iowa the pattern
was not the same as in Illinois. No land area corresponded
to southern Illinois in soil characteristics or farming
types. The maps also do not reveal an area of Iowa similar
to eastern Illinois that declined in importance after 1890.
The number of beef cattle in Iowa surpassed the number in
Illinois in 1880 and consequently the density in leading
Iowa feeding counties tended to be greater. Clinton and Muscatine counties were leading cattle feeding centers by 1870
and retained that position to the end of the century. West
Liberty in Muscatine County was the center of a region of
Shorthorn and Hereford cattle breeders and feeders by 1870,
while the surrounding counties of Cedar, Jackson, and Jones
had such residents also.[20]

From this eastern cluster, cattle feeding concentration
extended into east-central Iowa and into the southern and
western counties of heavy corn production. Counties in the
north-central area of the state extending south to Polk were
of most recent glaciation and had drainage problems that
hindered rapid farm settlement. They were grazing grounds

for cattle before the late 1880s; but after they were
drained, the farmers turned them to cash-grain combinations
with oats predominating.[21]

In the east north-central area wheat was the important
grain crop until the late 1870s, and farmers who raised cattle sold them at one or two years of age to feeders in central and southern Iowa. In 1877 J. H. Perry (Worth County)
wrote the _Prairie Farmer_ that "wheat has been the great staple here for years, and but little attention here is paid to
raising corn." In northeastern Iowa dairying became increasingly important toward the end of the century.[22] The northeastern counties contained some of the most broken and hilly
timberland of the state that served its owners better as
dairy pasture than cropland.

In the corn-producing counties of Iowa, even more than
in Illinois, it paid to feed corn rather than sell it because the transportation charges would take a good deal of
the profit. The eastern and southwestern sections of Iowa
came to be the epitome of Corn Belt farming where farmers
fed their abundance of corn to cattle and swine. So ideal was
Iowa for this crop-livestock economy that a publicity pam-
phlet of 1885 declared in effect, that stock feeding was di-
vinely inspired and fostered since "the God of Nature made
Iowa especially a corn-growing state . . . [and] it may
safely be assumed" that the surplus grain was meant for feeding cattle.[23]

There was less publicity about Iowa feeders than about
some of the widely known Illinois men who dealt in thousands
of head of cattle. Men such as William G. Hann (Clinton
County) and Oliver Mills (Cass County) introduced purebred
stock before the Civil War to improve the beef cattle. By the
end of the century others had followed their examples so that
both counties were leading beef feeding areas.[24] In Adair
County, which in 1870 had less than 1.5 beef cattle per hundred acres, farmers introduced improved stock in large numbers, and by 1880 the county had over 10.9 head per hundred
acres.[25] Alexander John (a farmer in Taylor County, which had
a concentration the same as Adair) started raising Shorthorns in the 1870s. By 1891 he had over 75 head of purebred
cattle, held annual sales, and exhibited only at the local
fairs. The 1890 census credited Taylor County with over 10.9
beef cattle per hundred acres.[26] In all of Iowa in 1879, only
441 farmers purchased or sold more than 300 cattle during
the year, but 12,679 bought and sold more than 20 head. Most

of these men resided in southwestern or east-central Iowa. Woodbury County had 17 large buyers (the largest number), but Muscatine County was the home of the largest dealer, Dennis Mahanah, who bought and sold 4,000 head during the year.27

George F. Green (Jackson County, Iowa, farmer) was perhaps typical of Illinois and Iowa feeders in the 1870s. He farmed 340 acres and handled from 100 to 200 head annually with the help of a son and two hired men. He bought small lots from local farmers in the spring and summer and occasionally bought droves as large as 50 to 75 head from a nearby auction dealer.28

The droving and grazing common to Illinois in the 1840s still existed in north-central and northwestern Iowa in the 1880s, where land in farms did not exceed half the acres in the counties in 1880. (See Map A.4.) Men such as A. D. Arthur of near Webster City purchased surplus cattle from farmers in the surrounding region to graze on prairie. In 1873 Arthur had 600 head and the next year 1,200 on grass in Hamilton County. In 1879 he pastured 1,200 in Wright County.29

General conclusions about location of feeders were difficult. Environment played a part, as seen in the limiting factors of Mason County and southern Illinois and the influence of wetlands on agricultural pursuits. But most of Illinois and Iowa produced corn and grass in abundance with no particular advantage of one area over another. The location of purebred raisers probably influenced cattle feeding by making improved stock available. Central Illinois and eastern Iowa had concentrations of breeders from an early period, which undoubtedly fostered beef cattle production. But why did the breeders settle in the areas they did? There is no adequate answer. Chance certainly was a factor as well as the desire to be in an area where relatives or friends had previously settled. Morgan and Sangamon counties probably impressed early Kentucky immigrants with the similarity to bluegrass country where cattle were well established. Other men came to Illinois and Iowa from the Ohio cattle feeding country looking for cheaper land on which to duplicate Ohio practices, perhaps on a larger scale. However, immigrants with no past experience in the business also took it up when they came to the cattle feeding areas in Illinois and Iowa. They imitated what seemed to be a good way to make money and probably succeeded as well as native-born Americans, depending upon their individual skills.

It is difficult to ascertain why certain cattle feeding areas developed as well as why certain areas sustained the industry over the years. New settlers imitated the Alexanders, Browns, and Strawns who were already in operation, and the readily available purebred cattle helped improve local beef quality in the face of range competition. But why then did few men imitate the Funks in McLean County, Benjamin Harris in Champaign County, or John D. Gillett in Logan County? Some reasons have been suggested that recognize the percentage of tenant farmers in the county or the value of the land in relation to cattle prices in the late 1880s and early 1890s. Many feeders felt they could not make money with low cattle prices and turned to production of cash grains, while others obviously made money and continued feeding cattle.

The amount of investment certainly played a role. In raising purebred cattle, a man needed considerable capital, but most farmers could get a start feeding a few head without too much trouble. Where they went from there depended more on the talents of the individual farmer than on the available capital. If a judgment is to be made, it must be in favor of personal factors accounting to a large degree for where cattle feeders located and the degree of success attained. Financial returns motivated many, but others participated because of the personal drive and satisfaction. In 1880 E. T. Brockway (Washington County, Iowa, feeder) wrote that he had "a passion for blooded stock," and though it might be easier to purchase feeder cattle, there was more personal "satisfaction" in raising cattle on his farm.[30] Personal skill in judging cattle by sight and touch and in the management of profitable operations were perhaps the main determinants in who and where feeders were. A reporter's description of Jacob Strawn: "The great passion of his life was and is—steers—cattle; to drive them, feed them, buy them, sell them; and then own other steers to go through the like process"; fits these cattlemen.[31] They had the drive to work, a good eye for cattle, a good head for figures, and pride in the finished product. Benjamin F. Harris had contests with his neighbors over who could produce the fattest cattle and then publicized his winning results.[32]

Cattle feeders were the elite of the farm world. Shiftless, lazy, or inefficient men could run a farm, but they could not succeed as cattle feeders. It took "a vigilant eye and a diligent hand" to fatten cattle.[33] Joseph McCoy said that "no right-thinking man can regard [cattle feeders] other than public benefactors."[34] If, claimed an Iowa speaker in 1876, he who made two blades of grass grow where one grew before was a benefactor of mankind, then the cattle breeders and feeders who added "an inch in thickness to the sirloin of a beast" were also.[35] In 1901 James ("Tama Jim") Wilson said that the "test of a farmer's ability comes when he handles live stock."[36] Indeed it did for many. The natural abundance of the Middle West led some settlers, economists, and historians to believe that anyone in the nineteenth century who had the capital could make a go of a farm. But the cattlemen knew better; they had a profound sense that the quality of the man influenced his success as an agriculturalist. The cattle feeder was a skilled craftsman, "a sculptor working not with such coarse tools as hammer, chisel, and brush, but with life forces, heredity, and the potencies of feeds."[37] In addition, there was a great pride and sense of personal satisfaction in feeding cattle. As J. H. Oakwood (Vermilion County, Illinois) put it: "Beside the gain, it is a great pleasure to a farmer of good taste, to care for and feed a herd of broad backed, round bodied, and beautifully formed cattle."[38]

The reverential deference paid to cattlemen was still apparent in the 1952 presidential address of Edward N. Wentworth to the Agricultural History Society. Wentworth (Armour's Livestock Bureau) recalled his first introduction to "higher agricultural motivations in 1900." As a cattle clerk at the Iowa State Fair he met the "master breeders" he had read about in the Breeders' Gazette, and it made him feel "as if I were seeing in a dream or vision the patriarchs of Genesis."[39] Though these men were probably not akin to Abraham, Isaac, and Jacob, it was nevertheless true that more was needed than merely capital and the idea of being a farmer to feed cattle successfully. In the cattlemen's view, skill, character, the intangible quality of the "eye," and the proper appreciation for one's duty to mankind were required to raise or fatten better cattle at a profit. As Wentworth stated, "the unselfish work of serving one's neighbor ministered to one's own selfish advancement."[40] A fortunate coincidence for many cattlemen.

C H A P T E R 8. AGE OF IMPROVEMENT I:
STATE AND PRIVATE INTERESTS

During the course of the century a large number of people, as individuals or as members of organizations, encouraged Illinois and Iowa cattle feeders with advice and practical information. In general, these people considered the
changes suggested to be improvements. Some advisers and
their ideas have been mentioned, but a number of groups or
institutions that championed "improvement" should be discussed because they functioned over most of the period from
1840 to 1900 and were a constant if not consistent force for
change. Chief among these were editors of agricultural newspapers; but state and local agricultural societies also
sponsored informational meetings and fairs, and private cattle breeders' associations encouraged change. (Federal institutions and the agricultural colleges are discussed in Chapter 9.)

John S. Wright (Chicago businessman, promoter, and member of the Union Agricultural Society of northeastern Illinois) published the first of the notable midwestern farm periodicals. In 1841 he started the Union Agriculturalist and
Western Prairie Farmer in Chicago. It became simply the
Prairie Farmer in 1843 and except for a brief period in
1858 and 1859 has retained the name to the present. Though at
first a monthly, from 1856 through 1900 (the years of development for the cattle industry) it was issued weekly. Although always a farm paper from the great livestock market
of the nation, the emphasis of the Prairie Farmer changed as
management changed, sometimes being very personal, showing
the editor's interests on practically every page and at
other times impersonal, with no identification of the writer.
The preeminent position of Chicago to the agricultural economy of the Middle West gave the Prairie Farmer an audience
far beyond northeast Illinois. Other Illinois farm newspa-

pers came and went. Some such as Emery's Journal of Agriculture (Chicago, January 1858 to December 1858) and the Illinois Farmer (Springfield, 1856-1864) merged into the Prairie Farmer. The Western Agriculturist of Quincy (1858) (which evolved into the Live Stock Journal of Chicago), the Drovers' Journal (1873), the Breeders' Gazette (1881), and the Orange Judd Farmer (1887) sustained publication through the end of the nineteenth century. Newspapers connected with a cause, including the "Granger" papers in the 1870s and 1880s, usually did not last, but they carried useful information in addition to their polemics.[1]

Mark Miller (Wisconsin and Iowa publicist) began the major farm newspaper in Iowa as the Iowa Homestead in 1862 when he consolidated the Northwestern Farmer and Horticultural Journal with the Iowa Farmer and Horticulturalist. Henry Wallace entered Iowa agricultural journalism in a small way in 1879, then edited the Iowa Homestead from 1883 until 1895, when he began editing Wallaces' Farm and Dairy renamed Wallaces' Farmer the next year.[2]

Concerned with the overemphasis on grain crops, editors wrote their own articles or borrowed from other papers in an attempt to influence farmers to pay more attention to cattle raising. "This idea of growing grains from a farm continually, with no stock to keep it in heart ought by this time to be exploded," the editor of the Prairie Farmer stated in 1850. "It will ruin a farm or a country which practices it. It is the continual taking out, without ever putting in anything. . . . In short, if a farm is to be kept up, it must be stocked with animals of some sort."[3] Whenever possible, periodicals stressed the value of cattle manure for fertilizer as an additional reason for raising stock. The farm journals also argued that cattle feeding offered a more profitable use for corn. Cattle prices did not fluctuate as much as grain prices, and to some this made cattle fattening seem a less risky and more dependable business; however, it required more capital investment than grain farming. The stock grower was more "likely to be a rich man at the end of twenty years than a grain grower. He may not be as lucky for a single year; but his gains are more steady and certain."[4]

In the last half of the century the Prairie Farmer carried numerous articles urging farmers to raise corn instead of wheat and to use the corn to fatten livestock.[5] Wood engravings, depicting improved cattle and notable fat cattle raised in the Middle West, accompanied articles (original

and reprinted) that urged improvement of breeds and discussed the points of good cattle and how to recognize them.
Often newspapers listed names and addresses of cattle importers or owners of fine herds in the state as well as carrying pages of livestock breeder advertising.[6] In the 1870s
and 1880s the quality of illustrations improved when life-
like engravings of good animals began to be used in the continuing campaign to convince farmers to work toward improvement of beef breeds. Writers agreed that it took just as
much to feed a poor beef animal as a good one and stressed
the point that profit was much greater in the good animal.
Breeders inserted more advertisements, and editors generally
mentioned notable cattle sales.[7]

Farm papers devoted space to discussions of care and
feeding of cattle and sometimes reprinted articles from
other papers or proceedings of English agricultural societies. Some of these articles conveyed little useful information to midwestern cattle feeders. One was a verbose treatise
on "flesh forming" and "heat producing principles" taken
from a paper read to the Royal Agricultural Improvement Society of Ireland. If a farmer read all the way through and
gleaned anything, it was probably that one should keep cattle warm in winter because they would use less of their food
ration to generate body heat and could use more of it for
growth.[8] That this came from England might have made it seem
more authoritative than a simple statement to the same effect in articles or letters from farmers, but probably its
original purpose was to fill out that issue of the paper.

More often, advice on feeding was offered in straight-
forward language and directed to feeders in Illinois and
Iowa. Articles and letters offered advice on how to make corn
fodder for feed and how to haul it from the field, warned of
the dangers of turning cattle into cornfields without sufficient water, and gave information on feed supplements. The
letter writers also offered rules on feeding: "always be
regular and systematic in feeding your stock, . . . neither
too little nor too much, too often nor too seldom,"[9] and instructions to compare costs and value received for various
feeds.[10]

As important as what and how to feed stock was the
type of animal to be fed. The Prairie Farmer instructed readers that the desirable steer for economical fattening would
have a sharp muzzle, broad chest, straight level back, prominent placid eye, and well-sprung ribs, all of which showed

a "disposition to fatten."[11] Farmers were warned not to be
fooled into thinking that offspring of fancy cattle shown
at the fairs would look that way naturally. The biggest and
best-combed animals got the most attention at fairs, but the
Prairie Farmer warned, "there is a point beyond which size
cannot be gained without the sacrifice of that which is more
important. Quality and size go together for a certain distance, but they part company after a while, and size then
takes to itself another ally, viz, coarseness."[12] Exhibiting
overfed cattle, the editors thought, did a disservice to the
breeders and to the farmers visiting fairs because men who
bought improved breeds would expect every calf to grow up to
look like the show animals they had seen.[13] The Prairie
Farmer summed up its feelings on stock management in an article that ridiculed "The Mismanager" by stressing the following points: breeding was significant, young stock should
be well cared for because they would never get over a poor
start, it cost as much to feed a good type of steer as a bad
type, and cattle needed winter shelter to be in good shape
in the spring.[14]

Though the farm papers were well intentioned, they did
not always succeed in providing pertinent practical information. Well into the 1870s the Prairie Farmer carried reports
of English Cattle Association meetings discussing feeding,
bean meal, mangels, trifolium, and tares, which had little
useful application to Illinois and Iowa farmers with their
fields of corn and bluegrass pasture.[15]

Similarly, newspaper columns were opened to controver-
sies based on theoretical principles rather than any practical experience with cattle feeding. One such argument in
the 1860s over "natural physiological principles" concerned
the use of salt. Some men argued that cattle had a "natural"
desire for salt, and others argued they did not. Some even
worked through a system of reasoning that proved that salt
was "never" beneficial to animals.[16] So much advice was offered on so many aspects of fattening cattle that one farmer
complained, "valuable as your paper is, yet it brings con-
fusion in one's head sometimes."[17]

Editors also had favorite cattle management practices
they continued to urge upon feeders without much success.
Such was the case with cattle barns. Editors and correspondents generalized from human experience and concluded that
cattle liked to live in buildings and would gain more weight
and be healthier if stabled, therefore it must be cheaper to

feed cattle in barns. "We have had too much out-door feeding and the time is not distant when we hope to see a change in this respect."[18] This attitude seldom took into account the cost of the barn, the extra cost of handling feed for stabled cattle, the gains made by hogs that followed cattle in a feedlot, and the fact that one man could feed more cattle in less time if they were in a lot rather than a barn.

For a number of years, in fact into the late 1870s, many editors advocated cooking cattle feed. Articles offered figures to prove it was cheaper to achieve a pound of gain with cooked feed, but their authors did not consider the increased amount of time spent in using this method of feeding. Other letter writers insisted that cooking food for cattle in addition to stabling them did not pay. Some articles on the subject were probably inspired as advertising for manufacturers of steam cookers. Regardless of the fact that corn and grass remained the major cattle feeds in price and economy of use, papers seriously discussed root crops and steamed forage as cattle feeds for the Middle West until the 1880s.[19]

The farm journals carried much solid material for those who really wanted information and practical illustrations of stock feeding operations and results. Feeders wrote about their experiences and gave their feeding plans. Articles cited experiments at the state agricultural colleges, and professors at agricultural schools occasionally wrote articles embodying the results of their experiments. Also regular county correspondents reported on current cattle feeding practices.[20]

When reporting on the controversy over Texas fever and other diseases, the Prairie Farmer tried to be impartial. But when pleuropneumonia spread westward in the 1880s and the magnitude of the disease control problem became more widely known, the Prairie Farmer editors strongly supported attempts to establish federal disease control measures. When attacked as alarmists for discussing the spread and possible consequences of pleuropneumonia, the editors retorted that their critics should help fight the disease, not deny its existence.[21] Other papers closer to the plains cattle interests insisted pleuropneumonia was a "fraud" and "racket" for the profit of veterinarians.[22]

From 1849 on, farm journals generally carried market reports giving Chicago receipts, shipments, and prices for different grades of cattle and often included marketing advice or a discussion of probable market conditions in the weeks or months ahead. Sometimes, especially in the 1850s, the market reports included the names of the men who fed shipments of good quality cattle. The journals also carried articles on local and state fairs and the American Fat Stock Show in Chicago (begun in 1878) as well as reporting on some of the meetings of cattlemen's associations, breeders' groups, and local farmers' institutes.

In keeping with an injunction printed across the front page: "Farmers Write for Your Paper," the Prairie Farmer carried much material in the form of articles, addresses given at meetings, and letters from farmers offering or seeking information on a particular subject. These letters and speeches called for and gave the results of the "experience of intelligent, close observing practical farmers," as a speaker at the Peoria County, Illinois, fair said in 1885. He noted that book learning was becoming respectable and was helpful, but there was no substitute for experience. "Knowledge is power," he said, "but this must be practical knowledge—a knowledge of the proper mode of doing things—not a mere vague notion of how they should be done."23 Farm papers offered themselves as a means of spreading this practical information about cattle feeding as well as other subjects. "We know that there is a good amount of enterprise, and no inconsiderable skill among [cattle feeders]. The question is cannot this skill be made available to others—cannot this spirit of enterprise be extended?"24 The agricultural press attempted to disseminate information on the skills and spirit of enterprise of the cattle feeder.

In the Prairie Farmer and the Iowa Homestead, space was also given to the dairy, veterinary, household, and editorial departments in addition to articles on other types of livestock and grain crops. Later in the last half of the century, the Breeders' Gazette and the National Livestock Journal catered more to livestock breeding and feeding but were not, from the feeder's standpoint, noticeably different in content from the Prairie Farmer and Iowa Homestead.

The source and content of letters to the Prairie Farmer and Iowa Homestead indicate that many read the papers and followed them for advice, but circulation figures give some quantitative, though by no means exact, evidence of their popularity and use. Farmers could and did pass papers around, so that only one reader per subscription is a minimum number. In 1880, the Prairie Farmer claimed a circulation of 12,000; in 1891, 35,000; and in 1900, 25,000. For the same three

years the Iowa Homestead claimed a circulation of 9,000,
10,000, and 40,075. In 1891 and 1900 Ayer's American Newspaper Annual credited the Drovers' Journal weekly edition
with a circulation of 18,200 and 14,310 and Breeders' Gazette with 12,000 and 22,721. Bardolph concluded from his
study, carried to 1870, that farm journals were more effective forces for improvement during the period than agricultural societies.[25]

In addition to periodicals, local and regional associations were important in promoting farm interests. They had
exhibitions and awarded prizes to farmers with good cattle
and held meetings to discuss matters of interest to cattle
feeders. The fairs and meetings were reported in newspapers
beyond the local area. In 1853, after several abortive attempts, leading Illinois and Iowa farmers organized state
agricultural societies as quasi-public bodies autonomous
from the state but governed by a legislative act and given
periodic financial support. The first major activity of
these societies was to organize an agricultural fair, which
they started in Illinois in the fall of 1853 and in Iowa a
year later. With the exceptions of 1862 and 1893 in Illinois,
both states had annual fairs from then on. The agricultural
press heartily supported these as a way to educate farmers.
There they could see the best representatives of fat animals
and excellent crops. The Prairie Farmer urged farmers: "Go
to the Fairs then for improvement—to see to question and to
reflect. It is the place to which you should take the best
of the herd, the flock, the field, the garden, and the
household. . . . Make them pay by what you may see and
learn."[26] Improvement was the watchword of the day, but farmers also liked entertainment; and fair committees soon found
that they could not draw good crowds with only fat stock
standing around. Despite the disgust of the Prairie Farmer
editors, ladies' equestrian exhibitions, horse races ("speed
trials"), and carnival booths made an early appearance and
stayed.[27]

Cattle shows relatively quickly turned into contests
among rather small groups of men in the pure-breeding business, with Shorthorns dominating fairs until the 1880s when
Hereford and Angus cattle became commoner. The Iowa State
Fair offered prizes, not restricted to purebreds, for fat
cattle each year to 1900, and Illinois did so until 1878
when this type was dropped for lack of interest. Purebreds
were to provide examples against which farmers could compare

91

their own cattle and toward which they could work. However, the beauty and bloodline interests of judges of purebred cattle ignored their merits for the beef market, and thus prizewinners could provide poor examples for cattle feeders. Also it was questionable as to how much fairs encouraged general cattle improvement throughout the state. Not many different breeders exhibited until the 1880s, and the·same familiar names won the prizes year after year; but the state fairs at least offered competition among some breeders and through publicity kept the subject of cattle improvement before the public.[28] This helped feeders by encouraging better cattle breeding within the state. Purebred bulls produced better grade steers that converted corn into beef more economically and matured at an earlier age than common cattle.

Agricultural society reports and transactions distributed throughout the state to county clerks, local agricultural societies, farmers' clubs, and political friends of the legislators carried extensive information on the fairs including lists of exhibitors and their addresses, prizes offered and awarded, and reports of the awarding committees. In addition the reports printed prizewinning essays on topics such as breeding and feeding cattle and included statistics on crops. By the 1880s, state societies issued regular crop reports that gave feeders some idea of the coming volume of the corn harvest. The magnitude of circulation and usefulness of this information was hard to estimate. It was directly available to some and formed the basis for newspaper articles that carried it to a wider audience. The legislature authorized only 3,000 copies of the first issue of the Illinois society's Transactions, 8,000 of the second volume, 3,000 of the third, and 7,000 to 10,000 of the others to 1870. Bardolph concluded that farmers valued them, though the volumes were hard to obtain.[29] The legislature limited the Iowa society's Report to 3,000 copies from its beginning to at least 1873. By 1900 both state agricultural societies had been reorganized. The Illinois State Agricultural Society was superseded in 1870 by the State Board of Agriculture, which administered the new State Department of Agriculture. The same thing happened in Iowa in 1900.

Of perhaps more importance for cattle improvement were the local agricultural societies that formed a loose confederation supporting the Illinois state organization. In Iowa, presidents of local societies elected the state society officers. Both states encouraged county agricultural societies by voting moderate subsidies to such groups. In 1860 Illinois had at least 68 and Iowa 36 local agricultural societies. By the end of the century Illinois had at least 72 organizations that sponsored fairs, and Iowa had over 101 county or district societies that qualified for state aid.[30]

The county and district societies functioned on much the same lines as the state groups. They sponsored fairs and held meetings where local men or occasional outside speakers read papers on topics of local interest or to encourage something new. For example, the Buel Institute meeting of October 1850 in Granville, Illinois, held a fair at which the number of cattle, though short of expectations, "were generally of a superior stock and reflected great credit on those who raised them." The society also heard an address by Arthur Bryant (noted horticulturalist from neighboring Princeton and brother of poet William Cullen Bryant) who discussed prominent faults of agriculture in that part of the state. One of the faults was the "too small amount of stock kept on most of our farms." "The beef in Illinois," he said, was "equal to any in the world," and he urged the farmers to raise more of it.[31] This was not an isolated sentiment. Across Illinois and Iowa in the 1840s and 1850s speakers at local meetings called for improvement of cattle or offered papers dealing with cattle feeds, or for example, "Winter Care of Stock."[32] A Bremer County man reported to the _Iowa Homestead_ in 1869 that the "farmers are waking up to the subject of improving their herds by the introduction of thoroughbreds," and that they had organized a farmers' club that winter with 50 members.[33] In the 1870s such groups as the Madison County, Illinois, Farmers' Club and the West Liberty, Iowa, Stock-Breeders' and Farmers' Convention heard papers by representatives of their respective state agricultural schools, discussing in detail the rearing and feeding of beef cattle according to the latest theories and experiments.[34]

In the local societies as well, breeders of purebred cattle tended to dominate the cattle divisions of the fairs as they dominated state fairs with their beauty contests for pedigreed animals. Too often men forgot that purebreds were a means to an end and not the end itself, and breeders were unhappy if nonpedigreed stock won prizes. In Scott County, Iowa, the agricultural society had no requirement limiting exhibition to purebreds in 1868, and the prize for the best herd went to a group with only one purebred. In reaction to

93

this, the writer of the Prairie Farmer article complained that a herd with almost all purebreds was passed over for the prize. The writer commented that fair committees would have to make rules eliminating nonpedigreed stock if breeders of purebreds were expected to exhibit.[35]

The Illinois State Board of Agriculture instituted the Fat Stock Show in Chicago in 1878 in order to have an exhibition in which cattle were judged on their merits as beef animals. The show was held each fall through 1891.[36] From the beginning, when judges were instructed to "award premiums to such animals as present the greatest weight in the smallest superfices—taking into consideration age, quality of flesh, and its distribution in the most valuable portions of the carcass," the Fat Stock Show was intended for the beef producers.[37] In time the show came to include dairy, poultry, and sheep and there were entries of purebred animals and established prizes for them, but only if they were shown as fat cattle intended for the butcher market. In that first year, five men (one each from Indiana and Kentucky, three from Illinois) entered 18 Shorthorns, two men (Will County, Illinois) entered 8 Herefords, three men (two from Illinois, one from Wisconsin) entered 5 Devons, and ten men (six from Illinois, one each from Iowa, Indiana, Kentucky, and Missouri) entered 53 grade or crossbreds. There were classes in each category for animals four years of age or over, three years and under four, two years and under three, and one year and under two. A group of yearling Shorthorn steers entered by James N. Brown's sons, which weighed from 1,275 to 1,480 pounds, had a claimed rate of gain from 1.9 to 2.2 pounds per day from birth.[38]

The Browns were leading the way in early maturity of beef and the Fat Stock Show helped make the idea popular. The Prairie Farmer observed of the Browns' winning yearling steers that the "quality of the flesh is very superior and much more ripe than in the average 3 and 4-year-old steers in our market, and proves conclusively that a better system of feeding for the more rapid growth and earlier maturity of young stock is imperative, if the greatest profit is obtained, and should be encouraged."[39]

The second Fat Stock Show offered prizes for carload lots and dressed carcasses. John D. Gillett won the prizes in both divisions. Gillett had been feeding fine fat cattle for many years, and the Fat Stock Show was an excellent place for him to show his skill. He had 54 of the total 95 grade or crossbred entries in the show. Gillett's winning carlots clearly showed the trend that came to be urged as the reason for early maturity. The yearlings averaged 1,313 pounds for a gain of 2.4 pounds per day, the two- to three-year-old carlot averaged 1,818 pounds for a gain of 1.8 pounds per day, and the lot over four years averaged 2,147 pounds for a gain of 1.34 pounds a day. As they grew older, the total gain became progressively less for the increased age, while progressively more food was required to keep the cattle gaining. Gillett won the carcass class with a three-year-old steer that dressed at 65.7% salable carcass to live-weight.[40]

For the 1880 show the state board decided that, in keeping with their aims of improving beef and encouraging early maturity, they would eliminate any classes for steers over four years of age. The Prairie Farmer endorsed the action with the time-worn phrase that it did not pay to feed old beef.[41] By the fourth show, though the majority of exhibitors came from Illinois, there were also entries from Canada, Kentucky, and Wisconsin.[42]

On the mundane level of financing, the Fat Stock Show continued to operate at a deficit, which caused some discussion of cancellation, but the president of the State Board of Agriculture felt that "the great benefit to accrue to farmers engaged in meat production by the continuance of this practical and important school of instruction is such as to make it necessary to continue the exhibitions. . . ."[43] The board continued to make up the deficit from state fair funds, and in 1882 they added a new classification—the cost-of-production competition. This was a needed addition, for some had complained that the wonderfully fattened animals that took the prizes were impractical for farmers to produce. The results from the cost-of-production competition would help settle the question. A comparison of costs in the first four years of the competition revealed that exhibitors had usually lost money on entries over two years old.[44] Thus the Fat Stock Show strove to be the "normal school for all interested in the production and consumption of meat."[45] "A Man of the Prairie" expressed a similar thought when he wrote, "I hope every stock raiser will do what I propose to do, get a copy of the State Board's report of this show, and study it for all it is worth. It should contain a great many important lessons for all of us."[46] The Prairie Farmer added its voice to the praise and said that the livestock interests of the country "owe a lasting debt of gratitude for the great work done through the Chicago Fat Stock Shows in elevating the standard of American meats, and for the resulting lessons in breeding and feeding for the market of the world."[47]

But not everyone was happy, especially the Hereford breeders and feeders who felt that the judges and management of the show were partial to Shorthorns. This was especially true after the state board put out a chart purporting to show "some value in determining the comparative feeding qualities as far as early maturity is concerned." The combined average weights and gains for all exhibits for four years appeared in a table showing that Shorthorns recorded the most rapid growth among purebreds; but because of the few Herefords shown during these years, there was no adequate representative base for a reliable comparison.[48] Hereford advocates specifically complained in 1881 and again in 1885 when the board received a long letter from C. M. Culbertson (Hereford breeder and feeder from Cook County). The letter cited "absurd" awards in past shows and complained that the show had originated as "a school, at which all might come to compete and learn." It was not serving this purpose because of poor judges, Culbertson complained.[49] T. L. Miller (pioneer Hereford breeder and long-time exhibitor at the state fair and Fat Stock Show) complained that the judges were butchers who only had experience with Shorthorns and their grades and favored them in judging. Better judges would recognize the merits of Herefords, according to Miller. He would not exhibit again until the board remedied the situation.[50] Miller and others were true to their words; the Prairie Farmer noted that the 1886 show was not as good in the beef department, as several noted feeders were "conspicuously absent."[51] However, a Hereford steer, which weighed 1,530 pounds with a carcass dressed at 67% liveweight, took first prize.[52]

Some breeders attempted to turn the Fat Stock Show into just another fair with prizes for purebreds only. In his report to the board, LaFayette Funk (of the well-known Illinois farm family and superintendent of the beef division of the show) said, "I believe the time is almost here when the exhibit of this show should be confined to the breeder. This idea is strongly urged by some of our exhibitors now." But in his address to the board, the next president discouraged any move along that line as "not in accordance with the principle upon which the show was founded."[53] Although strongly pressed for the next two years by various interests (Board of Trade, railroads, and Shorthorn breeders) to allow breeding cattle exhibits, the board repelled the purebred interests' attempt to subvert the purpose of the Fat Stock Show.[54]

The controversy over admission of purebreds, the establishment of a separate horse show, and the problems of finding adequate space for the exhibition combined to reduce public attendance and fat-cattle entries at the shows after 1889. There were no exhibitions in 1892, 1895, and 1896 and the Fat Stock Show came to a sad end in 1897. The manager (John A. Logan, Jr., son of the Illinois Civil War general who founded the Grand Army of the Republic) combined a society horse show, a fat-stock show, and a running battle with the board over management to produce a deficit of $30,705. At one time there were two janitorial staffs contending for custody of the door keys. From then on to the end of the century, the state board contented itself with sponsoring a ring for fat cattle at the state fair.[55] After much discussion as to what should replace the Fat Stock Show, the managers of the Chicago Union Stock Yards sponsored the International Live Stock Exhibition at the Stock Yards in 1900. With the first international exhibition, control of the show for fat stock passed to the livestock business interests in Chicago. This combination of packers, bankers, and railroads managed the show in terms of their interests in prime butcher beef regardless of breed. The prizes in the fat-cattle department of this first show went to Herefords and Aberdeen-Angus, with Shorthorns a poor third.[56]

The Fat Stock Show had also given exhibition space to beef by-products, which involved it in controversy and illustrated one of the problems of any attempt to achieve a unity of all agricultural interests. In 1885 the state board accepted the meat packers' offer to sponsor an exhibit of oleomargarine and brought down upon themselves the wrath of the dairy interests. That oleo made from processed beef fat was a legitimate product for a beef show seemed reasonable to the board, but the dairy producers, who exhibited at the Fat Stock Show also, looked upon oleo and butterine (leaf lard, milk, oleo oil, mixed with butter) as frauds and poisons in addition to being competition. The packers contended that oleo was a healthful food product so pure that "creameries" did "in fact use" it to manufacture what they sold for genuine butter, that packers sold a large part of their oleo to butter manufacturers, and that according to medical doctors oleo was a better product than low-grade butter. The dairy producers reacted within a few days. The Elgin, Ill., Board of Trade, representing the chief butter market of the nation, denounced with an unconscious pun, the exhibition of "bogus butter," as "not only in bad taste, but reprehensible, and merits the denunciation of every right minded man." Beef feeders responded with their own resolution backing the state board in encouraging every type of manufacture that aided the beef producers and stating that oleo was a legitimate food product and its manufacture of value to beef feeders, by giving more value to an otherwise waste product of their animals.[57]

In November 1885 the Ohio Dairymen's Protective Association resolved that the Illinois dairymen deserved sympathy for "the insult offered them, the surrendering of their rights and the betrayal of their trust by their State Board of Agriculture . . . in admitting bogus butter." The board, they charged, acted through either "imbecility or treason." The board stood its ground and refused to "enter into a conflict of preambles and resolutions with the Dairymen . . . nor . . . contest the palm for abusive billingsgate with the author of the [Ohio] resolutions; or [debate] the false assumptions and ill-natured criticisms and denunciations contained in the resolutions of the Elgin Board of Trade, and therefore content themselves with the following declaration of incontrovertable truths," asserting that it was not the duty of the board to suppress one form of food (oleo) for another competing variety (butter), that butterine and oleo when properly labeled and sold were not bogus but cheap wholesome substitutes for butter, and that the board was in favor of the farmer and laborer being able to get the best food for the least cost.[58] Illinois and Iowa beef feeders had a vital interest in the oleo trade because the more beef by-products packers could use, the more beef cattle were worth. The controversy went on in the country (and the world at large too) as dairymen the nation over attacked oleo and butterine retailed by the meat packers, but oleo continued to be exhibited at the Fat Stock Show and the dairy exhibitors tolerated it "instead of being there with the tomahawk and scalping knife" for which the board commended them. The dairy interests were not politically strong enough to effectively attack oleo until 1902.[59]

The Fat Stock Show offered an index to the changes
made by the leading beef feeders of the Middle West. The
winning cattle in the show forecast and documented the trend
toward early maturity. Early shows offered a category for
four years and older but soon dropped it, and in 1891 the
show eliminated the rings for cattle three years old and under four, and led the way toward baby beef. By the early
twentieth century two-year-olds dominated the markets and
prime yearlings (baby beef) were encouraged, though by no
means did they represent a large portion of the stock marketed. One commission firm estimated that baby beef represented only 1% of the fat cattle they marketed.[60] By the end
of the century, however, leading Illinois and Iowa cattle
feeders had completed the shift toward early maturity they
had begun in response to western range competition. Though
not accounting for a majority of cattle receipts in Chicago,
Illinois and Iowa feeders did lead the industry in marketing
at early maturity and in turning good range stock into prime
beef in their feedlots.

The two basic requirements for early maturity were
physical improvement of beef cattle by purebreeding and
changes in feeding practices to accelerate weight gain.
Though it came into prominence at the time of western range
cattle competition, improvement of native cattle with purebred crosses had been going on for some time, though not as
fast as many wanted. Fair associations and the agricultural
press stressed the importance of "breeding up" stock by using quality bulls, but it was not always an easy matter to
convince "our old fashioned farmers. . . , that any profits
could be gained by costly and mixed blood." Some who experimented failed to do the necessary fencing to keep out stray
bulls in order to maintain the benefits of having bred improved stock.[61] An Iowa farmer complained that it was akin
to the "tulip mania" to pay high prices for purebred imported bulls, but the Prairie Farmer replied that of course
the bull as beef was not worth very much but the gain came
in the quality of the offspring and the better prices for
them as beef, and the paper continued to urge improvement.[62]

Some men such as Oliver Mills (Cass County, Iowa) knew
the value of improved stock and did not need urging to utilize it. In the late 1850s Mills planned a cattle feeding
operation on public range. Because he expected to purchase
a large proportion of his cattle in the area, Mills distributed twelve purebred Shorthorn bulls free for local

farmers to use in producing good grade stock, which he could buy and fatten on grass.[63]

Purebred Shorthorn bulls were a major element in beef cattle improvement. Though there were some Devons in Illinois and Iowa, they were a minor influence in improving beef cattle.[64] Men brought varieties of Shorthorn cattle to Ohio and Kentucky as early as 1800. An importation of English Shorthorns to Kentucky in 1814 established the bluegrass region as a major breeding center. Kentucky men brought Shorthorns into Illinois in the early 1830s and, in the next two decades Shorthorn importing companies or associations operated in Ohio, Kentucky, Illinois, and Iowa. From 1833 to 1855 Illinois men bought at least 118 registered Shorthorns calved in Kentucky or Ohio and 11 from New York and had produced 82 registered Shorthorns on their own farms. Iowa men had bought 33 Shorthorns from Kentucky, Ohio, or Illinois in the same period.[65] In 1855, 57 registered Shorthorns were calved in Illinois or imported. Most were found in the west-central counties of Morgan and Sangamon or around the fringes of Chicago. In 1855 only six Shorthorns were calved in Iowa or imported, and they were found in the extreme southeast in Lee County.[66]

Most of the breed improvements were individual projects in the early years. Some failed, as when a Colonel Oakley took a purebred Shorthorn bull and two cows into Tazewell County, Illinois. Farmers did not take care of them or make an effort to be selective in their crossbreeding. The superior blood lines were downgraded and lost.[67] On the other hand, by 1873 John Scott (prominent Iowa farmer) cited the fine grade of cattle in Cass County as evidence of the success of Oliver Mills's distribution of purebreds.[68]

Prominent central Illinois cattle producers formed the Illinois Stock Importing Association in 1857 to bring purebred cattle from England. In August 1857 the association sold its first shipment of 20 cows and 7 Shorthorn bulls at auction in Springfield. Nine of the cattle were sold to Sangamon County men; four and two were sold to men in the adjoining counties of Menard and Morgan respectively. Three each were sold to men in St. Clair and Will counties and one each to men in Cass, Champaign, Jersey, Madison, Pike, and Scott counties.[69]

By the middle 1850s "Long John" Wentworth (Chicago businessman and politician) made imported cattle available through his Illinois Breeding Association, and in 1857 the short-lived Ohio Stock Raising Company drove 60 purebred Shorthorns, which were soon dispersed, to Butler County, Iowa.[70]

By 1860 considerable concentrations of purebred cattle existed in Illinois and some in Iowa that exerted a local influence at least, but reports from county fairs continued to show few purebred entries and a very small number of different owners. Generally, less than a dozen different men won the Illinois State Fair prizes for Shorthorns, and they were the same men year after year.[71]

After the Civil War ownership of Shorthorns spread further. From 1855 to 1870, Illinois men bought at least 672 registered Shorthorns from Kentucky, Ohio, or New England and produced 2,445 on their own farms. Iowa men bought at least 226 Shorthorns from Kentucky, Ohio, New England, and Illinois and produced 210 on their own farms. However, in the last quarter of the century, Hereford cattle began to gain in favor as superior beef animals. Men found Hereford bulls especially suited for improving range cattle.[72] (See Maps A.13 and A.14.)

Breeders contended with each other at the fairs and in the press as to which breed was best for beef. In 1880 C. L. Hostetter (Carroll County, Illinois) wrote that he bred Shorthorns because they made the "best beef" while at the same time T. L. Miller (Will County, Illinois, breeder) reported that he kept Herefords because they were "pre-eminently the beef cattle of the world."[73] The farm periodicals urged cattle improvement through articles, editorials, and advertisements. Indeed the editor of the Iowa Homestead declared: "The advantage of improved blood in stock can be so satisfactorily proved that we shall not dwell on this point further."[74] From fairs, farmer associations, the Department of Agriculture, and the agricultural colleges also came urgings to improve beef cattle. By the last quarter of the century breeders had formed their own associations. In 1881 in Chicago prominent Illinois and Iowa breeders were the major force in creating both the American Hereford Cattle Breeders' Association and the American Short-Horn Breeders' Association, and in the next few years they set up state associations. The Iowa Shorthorn Association reported at least 142 breeders in 58 counties by 1883.[75] Muscatine County was a center of Shorthorns in Iowa. In the vicinity of West Liberty, there were 22 breeders with over 700 cattle.[76] Much of the breeder organization was beneficial, but there were also

101

some disadvantages—financial speculation being one. Short-horns suffered from this particularly, as men bid up the price of desired pedigree lines and priced Shorthorns out of the market. By the end of the century men bought Hereford and Aberdeen Angus bulls in preference to the Shorthorn breed which lost its commanding position in numbers of purebred cattle.[77]

There were also nonpartisan societies that included men interested in any kind of improved stock. The Iowa Improved Stock Breeders' Association, founded in 1875, held meetings that included presentation of papers on aspects of cattle breeding or discussions of cattle diseases, types of pasture, and the value of farmers' institutes.[78] Various Illinois breeder groups combined to form the Illinois Live Stock Breeders' Association in 1895. Its annual meetings had some programs devoted to cattle feeder problems and presented by Dean W. A. Henry of the College of Agriculture, University of Wisconsin, and Henry C. Wallace, editor, Iowa farm expert, and journalist. University of Illinois professors presented papers on such subjects as corn and cattle, feeding improved stock for the butcher, and the fundamentals of breeding.[79]

It is difficult to estimate how fast farmers improved their stock. Purebred cattle were available in Illinois and Iowa generally after the Civil War and many commented on the improved condition of cattle, but according to the census of 1890, 74% of Illinois beef cattle and 68.8% of those in Iowa still had less than half their bloodlines from improved cattle.[80] Even so, improvement of cattle was noticed in the beef market. In 1870 the Prairie Farmer remarked favorably on the increasing number of quarter-, half-, and full-blooded Short-horn steers coming to market from places that formerly had sent only native steers. An Iowa man commented that his steers were of better quality at a younger age after years of crossing them with Shorthorns.[81] County and state reports in the 1870s and 1880s noted improvements in cattle ascribed to the proximity of local breeders or the fairs, which gave men a chance to see good stock.[82] Farmer "S" reported from Powe-shiek County, Iowa, in 1876 that "the number of fine cattle that have been brought into our county the last year makes some of us think that we will soon be up with any county in the state in raising No. 1 cattle for breeders and beef."[83] In Knox County, Illinois, that same year "L.S.J." said they had "a good feeling . . . in regard to the improved breeds of cattle."[84]

But large-scale improvement was a slow process, and the
Prairie Farmer and others concerned called for greater effort for improving beef cattle "even if it may appear like
useless repetition . . . to employ further argument to show
the value of such improved over common stock." A ride
through parts of Illinois disabused the editors of the
Prairie Farmer of the idea that it was useless repetition
because they saw evidence of a great need to "impress the
facts so often presented in these columns on the minds of
men who are not abreast with the advance of the age in this
matter."[85] The Iowa Homestead urged farmers to upgrade
their stock in 1869 and was still stressing the subject in
1890.[86] The editors urged men to visit fairs and attend cattle sales to see and buy good stock.[87] The markets of the
world demanded better beef than common cattle produced and
the day had passed when common cattle profited the feeder,
wrote an Iowa farmer in 1880.[88]

The quality of cattle in Iowa improved more slowly than
in Illinois, and there were not a large number of purebreds
in the state until the 1880s. One reason for the slow pace
of beef cattle improvement was the complaint by many farmers
that the purebred bulls showed more fineness and style than
they did satisfactory offspring. Breeders of Shorthorns, especially, became too interested in fine points of pedigree
and show style in their cattle rather than the ability to
sire beef animals.[89]

Advocates of improvement claimed that a good improved
steer on the same feed would outweigh a common steer by 300
to 500 pounds at two years of age. Improved steers also
brought a better price because of the quality of the meat.
They simply were more efficient at turning grass and grain
into beef.[90] Though the Prairie Farmer reported in 1879 that
only one-fourth of the steers received in Chicago showed
"good blood" and continued to complain of the abundance of
poor cattle coming to market, increasing evidence of the improvement campaign showed up in the market and at the Fat
Stock Show.[91]

In 1878, the best steer at the first Fat Stock Show was
a grade Shorthorn, three years and seven months old, that
weighed 2,185 pounds. The steer was nearly the "model of
perfection for a choice butcher's bullock, that of an oblong
square." The awarding committee described it thus: "The back
of this steer was straight and broad from shoulder to loin,
with flesh deep and even as a cushion; the ribs were well

back, long and well covered, wide and deep chest; shoulders were well rounded, with a neat and short neck; head small, with fine expression. The hind quarters were loaded with flesh and free from bunches of inferior meat. This animal, combining so many good points in body, had short, fine and well tapered limbs, thereby giving the greater profit to the breeder and consumer, and the least offal to the butcher."[92]

Even then, Gillett was in the process of changing feeding practices. Previously, he had sold fat cattle at four years of age and at weights close to 2,500 pounds. A Scottish visitor who described Gillett's operations in the early 1880s ("the grandest bestial display I have ever seen in one man's possession") noted that Gillett was preparing to sell two-year-olds fattened for the export market.[93] In 1879, Gillett's prizewinning entry in the yearling class of the Fat Stock Show weighed 1,196 pounds with an average gain of 1.97 pounds a day since birth. This grade Shorthorn steer was described as "almost a prodigy in the nature of the species."[94] But it was hardly a prodigy; other entries in the Fat Stock Show revealed equal attention to the possibilities of early maturity. From 1878 to 1885 feeders entered yearling grade steers averaging from 1,000 to 1,500 pounds with average rates of gain from 2.17 to 2.97 pounds a day from birth. From 1882 to 1885 entries in the cost-of-production category for 24-month-old grade steers had average weights from 1,370 to 1,573 pounds, average costs per hundred pounds of gain from $4.42 to $4.78, and net profits from $18.61 to $20.59 per head. Similarly cattle at 36 months for the same period averaged 2,156 pounds at a cost of $6.89 per hundred pounds and a net loss of $21.70.[95]

By 1885, in the category for yearling dressed carcasses, liveweights averaged 1,232 pounds at a gain per day of 2.20 pounds and dressed carcasses averaged 65% of liveweight. From 1880 to 1885 all dressed carcasses entered averaged over 60% of liveweight.[96] Also by 1885 Shorthorns had begun to yield their monopoly of prizes to Herefords and Aberdeen Angus purebreds and grades, and there was lively competition and debate over prize awards to the various breeds.

As early as 1880 the Prairie Farmer favorably commented on "baby beef" and the examples at the Fat Stock Show, but not all were pleased. Critics of that kind of "progress" acknowledged that two-year-old cattle furnished a large portion of beef marketed but contended that some of the "so-called improvement" was a forced "unnatural degree of fat-ness." Attacks on early maturity drew the answer that it was silly to insist that only four- or five-year-old beef tasted good. Proponents of early maturity in beef cattle said it was an age of improvement and beef was only one of the elements that had benefited. They pointed out that improved varieties of fruits flowered sooner than they had 20 years before, and with no lack in quality. Young animals kept on full feed for their short life, it was asserted, would have better marbled fat rather than a layer of meat enclosed by fat put on at the age of three or four. The carcass class at the Fat Stock Show proved that good beef could be produced and marketed at a profit at two years. This was by no means considered the ultimate in early maturity because in 1885 the National Livestock Journal predicted that men would soon be selling 1,400-pound steers at 20 months of age.[97] The Breeders' Gazette concluded that the American public had agreed on one thing at least in 1886: they wanted tender beef in contrast to "something to bite at" which the British still seemed to prefer.[98]

Feeders found it paid to sell beef at a younger age. The quicker turnover reduced the risk of capital loss, the first two years of feeding put on proportionately more meat for the cost, and the same amount of capital could finance a greater volume of business by reducing the age spread of animals maintained on farms. By the 1890s many considered that a three-year-old steer was a "back number, and is not wanted in the feed lot," and feeders were producing first-class beef at 20 months.[99] Illinois and Iowa men who continued to feed three- and four-year-olds got "precious little encouragement" from the market because their cattle sold for the same price per pound as two-year-olds.[100] A number of sources commented on the fact that younger cattle dominated the market by 1900. The census reported that cattle went to slaughter at an earlier age because of improved breeding and feeding; and a commission firm summarized the December 1900 market as one showing a preference for 1,300- to 1,400-pound steers. "The heavy cuts are out of style and all classes of buyers would prefer lighter beeves." The drive for early maturity through cattle improvement had paid off.[101]

CHAPTER 9. AGE OF IMPROVEMENT II:
FEDERAL INFLUENCE

The federal government aided farmers in general and
beef producers in particular in a number of ways. This aid
began inconspicuously in 1837 when the Patent Office started
distributing seed of new or improved plants. That office issued its first report on agriculture in 1841, and by 1849
the report comprised a separate volume of the Patent Office
Reports. In 1862 Congress established the Department of Agriculture and added divisions to it from time to time. Most
important for the beef producers was the Bureau of Animal
Industry (BAI), created in 1884 to supervise the veterinary
service founded the year before. In 1890 the Federal Meat
Inspection Service was added to the bureau.[1] Until the late
1870s and early 1880s, when the department began issuing a
variety of publications, the annual report was the major ve-
hicle for providing information to the public. The early
printed reports were sent to state agricultural societies
and distributed by congressmen to farmers and to newspapers,
which frequently extracted items of interest so that information reached a wider audience than might be expected.
The Patent Office Reports frequently printed letters
from selected correspondents scattered through the states
and contained articles on specific subjects. The 1844 report
contained reprinted articles on the amount of "fat-forming
principle" in certain foods and the contemporary advice
about cooked feed. Much influenced by British writings the
editors for the Patent Office urged farmers to feed carrots,
rutabagas, mangel-wurzels, and corn fodder to their cattle
during the winter.[2] Though much of the information was impractical, the Patent Office had good intentions and urged
farmers to raise more cattle to utilize grain and yet get
manure to help maintain soil fertility. In the effort to
foster beef cattle improvement by upgrading cattle through

106

the use of purebred sires, the reports carried articles on types of purebred cattle, although there was emphasis on Shorthorns. One such article concluded that the great hope of the U.S. beef raisers lay in good grade cattle that "are thrifty, and lay on flesh rapidly and evenly, and are ready to turn off at two and a half or three years old."[3] The Department of Agriculture, when it superseded the agricultural branch of the Patent Office in 1862, issued similar reports. The new series of reports contained articles supporting cattle improvement, giving feeding instructions and the number of cattle in states, and analyzing various feeds. By the middle 1870s the reports began to rely less on European experiments for information and to print results from more pertinent sources. The 1873 report carried the results from an experiment at the Illinois Industrial University that concluded it was not advisable to cook feed for steers.[4]

The department reported on its own technical and scientific investigations. In 1880, for example, there was an article by department chemists on an analysis of the juices of cornstalks.[5] The department also began to print statistical reports on the number of cattle in the country, the volume of trade in live and dressed beef, market receipts, and price ranges; but these did not provide the kind of statistics that would help feeders in immediate planning, although they could appreciate the size and complexity of the market. In general, before 1880, reports of the Patent Office and Department of Agriculture were of more use to later historians than contemporary farmers. Limited circulation and often impractical information reduced their usefulness; but the printed correspondence indicated what farmers were doing, and agricultural editors gave wider dissemination to some information by borrowing from the reports to fill their columns.

Of more importance, in the 1880s and 1890s the department started to issue special series of publications. One bulletin discussed early maturity, the economy of feeding, the importance of good breeding, and how to select feeder cattle. There were illustrations of good and bad types and statistics from experiments to illustrate with specifics the points made about age and feed.[6]

The BAI played the most important part in government aid to beef feeders. Its publications dealt specifically with beef cattle problems such as correlations of estimates of numbers with estimates of population growth and price cycles or specific discussions of the "Livestock and Meat Traffic of Chicago" that would tell a feeder just how the market worked, what type of cattle the packers wanted for different purposes, and what influenced prices. In one report, Professor Morrow of the University of Illinois discussed cattle rearing and feeding in the Middle West; in another report a department inspector discussed the cattle business in Illinois with specifics on type and amount of feed, ages and weights of feeder cattle, and estimates of costs, to which, he concluded, not enough feeders gave careful thought. These were reports of practical value to men who had access to them and could see just what the trends were in cattle production in the Middle West. The publications became more specific and producer oriented soon after the turn of the century with such items as Bureau Circular No. 105 describing "Baby Beef" and telling how to raise it.[7]

Disease control was the most important of all services of the BAI and of major significance to beef producers as well as all other livestock interests in the nation. Originally begun in 1883 the veterinary division of the department was incorporated into the Bureau of Animal Industry in 1884. In addition to the long fight against Texas fever, the bureau played a significant part in eliminating pleuropneumonia, the other major threat to midwestern beef cattle in the last half of the nineteenth century. Much of the effort of the Department of Agriculture had been devoted to investigations of animal diseases, but it was not until the 1880s that public pressure and the alarming spread of pleuropneumonia persuaded Congress to give the department power to do something about the situation. In addition to the danger to American beef cattle, pleuropneumonia had been the justification for the British Order-in-Council of 1879 requiring slaughter at the port of entry of American live cattle sent to England.

Introduced into the United States in 1843, pleuropneumonia spread from New York and reached Massachusetts in 1859. The Massachusetts government by prompt and forceful action checked and eliminated the disease within the state by a quarantine and then forced slaughter of all exposed animals, but no other state took similar action. The Illinois State Agricultural Society expressed concern about the disease, expecially the Massachusetts quarantine, and in 1859 appointed an agent to investigate the situation. The agent's report to the society confirmed the existence of the disease in Massachusetts, but said it did not exist in Illinois. Pleuropneumonia continued to spread in the East outside Massachusetts and was very much in the news in the 1860s. A second report to the Illinois Society in 1866 concluded that the only remedy was quarantine and slaughter.[8] One of the most valuable results of the Department of Agriculture's statistical activities from 1866 to 1877 was proof of the magnitude of animal disease and its alarming spread. In 1877 Commissioner of Agriculture Le Duc delivered a massive report on diseases to the Senate and in effect started legislative machinery moving toward setting up the BAI. Despite strenuous objections of certain livestock interests in the South and Southwest; of packers; and of commission firms who argued that a federal disease control program violated states rights, that ignorant veterinarians would needlessly slaughter cattle at a great loss to the industry, and that pleuropneumonia was not contagious, Congress created the BAI. The combination of support generated by the reaction to the British Order-in-Council, by an outbreak of pleuropneumonia in Illinois, and by the success of Wyoming ranchers in convincing eastern investors that unchecked pleuropneumonia on the plains could literally kill off dividends and capital investment overwhelmed the last-minute personal lobbying efforts of Nelson Morris for the packers. The new bureau immediately found the disease in Ohio and Kentucky also. The first law gave the bureau only quarantine powers without authority to slaughter diseased cattle; but that was remedied in 1886, and from then on there was swift action against pleuropneumonia though not without much opposition. Because of the insistence that slaughter was the only effective method of control, many attacked the bureau in and out of Congress. The director, Doctor D. E. Salmon, one of the great men in American veterinary history, was labeled incompetent by the minority report of a House of Representatives committee, in spite of the fact that the second Illinois outbreak in 1887 and department tests proved that inoculated and apparently cured cattle could and did carry the disease to reinfect the area.[9]

The BAI was viciously abused by some of the agricultural press for its work that hurt short-term profits. Field and Farm, representing Colorado interests, charged that the pleuropneumonia scare was a "fraud" by "racket workers," "quack veterinarians," or "designing cranks" as a way of getting federal money for "riotous living." The editors continually maintained that pleuropneumonia did not exist and when federal appropriations ceased all the "sick" cattle would get well immediately, and in the meantime the scare hurt the cattle business. But the BAI pressed its work and within five years eradicated pleuropneumonia from the United States.[10]

At the same time that the BAI worked to control Texas fever and pleuropneumonia, it investigated foot-and-mouth disease, anthrax, and rinderpest, none of which was rampant but did pose threats, as well as blackleg, which was a problem for improved fat calves and yearlings. Blackleg came from a microorganism spread by spores that survived in the soil of a pasture for years, which accounted for its recurrence in areas for no apparent reason. The organism entered the animal through the mouth or some puncture wound in the skin or foot, grew only in the absence of oxygen, and was fatal within 36 hours. The French discovered a vaccine for blackleg, and the BAI distributed free vaccine in the United States from 1897 to 1922. Before the use of vaccine, the department estimated that 20% of calves in infected areas died, though often owners attributed the demise to anthrax or clover poisoning.[11] With such a high mortality rate, blackleg was a serious problem to cattle raisers in certain areas of Illinois and Iowa. A 20% loss could eliminate any profit to a cattle feeder for the year.

The BAI also tried to help cattle feeders by taking steps to eliminate foreign criticism of dressed beef. Through bureau efforts Congress passed a law requiring federal inspection of slaughtered dressed beef. The Meat Inspection Service started in 1890, and in 1893 inspectors passed as disease-free and tagged 1,036,809 quarters of beef for export and 10,534,102 for interstate trade; but because of lack of funds the inspectors could cover only the major packinghouses and stockyards. Congress did not pass a truly effective inspection law until 1906 after the controversy stirred up by Upton Sinclair's novel The Jungle, which attacked the packers in general and Armour in particular.[12]

The federal government had also aided the beef feeders indirectly with the Morrill Act in 1862, which established the land-grant agricultural colleges. In the latter quarter of the century, the government and the states combined with the agricultural colleges to create another institution of improvement, the agricultural experiment station. Here was an organization within each state at last able to conduct practical experiments in an attempt to find answers to questions that breeders and feeders had been debating for years concerning feeding rations, shelter, and costs. Authorized by the Hatch Act of 1887, the Illinois and Iowa agricultural experiment stations embarked upon significant roles in improving the agricultural practices of their states.[13]

For many years spokesmen for farm interests had demanded some practical aid to agriculture. In the 1850s, they had called for agricultural colleges or experimental farms to educate farm sons and daughters "in the science, beauty, and goodness of their occupation."[14] The Civil War interrupted the movement, but Illinois had an Industrial University and Iowa a State Agricultural College in the late 1860s bolstered by the "awakening . . . conviction in the minds of very many of our farmers, that our profession is not merely one of plodding drudgery, but that it was also conceived, born and had its being in a science having for its study and investigation the broadest expanse of domain and deepest research."[15] The universities pursued some experiments in the 1870s and 1880s that were discussed in reports to the legislatures, newspapers, or publications of the state agricultural societies. E. S. Lawrence (head farmer of the Illinois Industrial University) commented on rearing and feeding cattle in the Fifth Annual Circular of the university,[16] and the Prairie Farmer carried reports of university feeding experiments in 1874 and 1875. These two experiments concluded that beets did not make a good fattening food for steers and "that the best and cheapest mode of feeding is to feed shock-corn in a sheltered yard." However, the major interests of the scientific investigators at the Illinois and Iowa institutions were horticultural and grain problems. Again, farmer conventions called for federal support for experiment stations, and the National Livestock Journal, in demanding their establishment, asked why it was still an unsolved problem as to how much grain supplement should be fed when cattle were on grass.[17]

Part of the problem was in the conflict raging over the purpose of the Morrill Act, the expectations of farm spokesmen, the realities of the agricultural college budgets, and faculty aims. In both states, faculties had developed a general education program with most quality instruction in the sciences; they had not limited the program to vocational agriculture as many economy-minded legislators, agricultural editors, and farmer spokesmen wanted. However, after some competent work, the agricultural departments of both schools declined in the 1880s to only a handful of students and a meager staff. The agrarians attacked Iowa State College in the press and legislature and forced a change, although they did not get humanities and social sciences banned from the curriculum. The new administration in 1891 separated the experiment station management from the old professors, placed a leading critic, James ("Tama Jim") Wilson, in charge, and added new staff that revitalized the agricultural work. The University of Illinois suffered a similar crisis and reorganization of the agricultural work in 1894. After that the quality of the work, both agricultural and scientific, improved.[18]

The government had helped resolve the crisis in agricultural education by providing funds for expanded work through the Hatch Act of 1887 and a supplementary appropriation act the next year, which provided $15,000 annually to each state to operate an agricultural experiment station as part of the university. Illinois and Iowa established stations in 1888. Most of the Illinois station work concerned dairy, fruit, and field crops with important work done on corn. Of the 11 annual reports, 60 bulletins, and 17 circulars issued by the station from 1888 to 1900, only Bulletins 2, 9, 36, 43, and 46 directly concerned cattle feeding experiments. Other experiments on alfalfa, cowpeas, and soybeans contributed to the store of knowledge about cattle feeds in comparison to corn fodder and ensilage. Bulletin 26 in 1894 was perhaps the most important for interested feeders, as it tabulated 180 replies to 250 letters sent to Illinois livestock feeders soliciting information on types of livestock, kinds of feeds, length of feeding period, and evaluation of feeding as a business. Here was a cross section of opinion on the latest practices by the leading beef feeders for anyone who read the bulletin. Interest in beef experiments increased at the experiment stations after the turn of the century, and the legislature added significantly to federal funds to allow larger scale operations.[19]

The Iowa experiment station did more of interest to beef feeders than the Illinois station, including five beef cattle experiments and three related feedstuff investigations in its 43 bulletins and 11 annual reports to 1900.[20] Bulletin 2 offered chemical analysis of fodders but left the reader to draw his own conclusions about what it meant; Bulletin 6 concluded that feeding shelled corn to steers produced more gain cheaper than corn and cob meal did. One of the more useful, Bulletin 9 listed various feedstuffs with an analysis by percentage of digestible nutrients and offered instructions on choosing rations for different feeding purposes depending upon food cost and availability. Bulletins 24 and 33 dealt with comparisons of steers and heifers and concluded that it did not make any difference whether heifers raised for beef were spayed, that heifers made good beef as economically as steers, and that packers were not justified in the price discrimination against heifers. The experiment followed two different groups from beginning feeding to finished slaughter and shipment to retailers. This particular experiment received notice in the Butchers' Advocate, which printed a summary of the experiments as evidence of packer profiteering by use of price discrimination against heifers.21

In 1898 the Iowa station experience produced a general but short survey of cattle feeding in the form of U.S. Department of Agriculture Farmer's Bulletin 71, "Some Essentials of Beef Production," by the Iowa station director, Charles F. Curtiss. However, both beef states lost their laurels to Wisconsin for the first major book on feeding, by Wisconsin experiment station director and dean of agriculture at the university, W. A. Henry, who in 1898 published Feeds and Feeding: A Hand-Book for the Student and Stockman. Herbert W. Mumford, professor at the University of Illinois, did not write his Beef Production until 1907.

Experiment station bulletins and university circulars did not meet all the demands of the improvers. Men had to obtain copies of the publications and sometimes puzzle out the conclusions without being able to ask questions. It would be better yet, many thought, if the professors went to the farmers in person and gave talks and answered questions about practical farming problems. The regents of the university in Illinois had sponsored lecture meetings in various parts of the state in 1870 and 1871. Members of the faculty spoke on agriculture in general and then on specific topics, one of which usually had something to do with beef cattle. In 1873 there were eight weekly meetings around the state; but from then on to 1880, meetings (now called farmers' institutes) took place only at the university in Urbana.22

In 1882 the State Board of Agriculture started sponsoring institutes in each congressional district even though some failed for lack of attendance, and by 1889 it sponsored 9 district and 33 county institutes. That same year, the state appropriated $100 to each congressional district for institute purposes. From 1891 to 1897, the legislature provided each county institute with $50, and in 1895 created a statewide organization, the Illinois Farmers' Institute, to coordinate county activities and with authority to publish 10,000 annual reports a year. From 1897 the state institute had an annual appropriation of $7,000. It lasted well into the twentieth century.[23] The meetings stressed "good farming practices" such as urging men to keep livestock in order to feed their corn rather than sell it and discussed the need for early maturity and quality in order to compete with grade range cattle. The central institute attempted to co-ordinate activities, to supply speakers to the county meetings, and to encourage the submission of essays for its Annual Report. However, most of the county meetings consisted, for example, of the farmers gathering to hear a local minister invoke God's blessing, an address of welcome and response in "poetry," and a "song by Miss Callie Fisher" and to read papers to each other about rotation of crops, cattle feeding, and other topics.[24]

In Iowa, the State Agricultural College initiated institutes in 1870 and held them irregularly over the next decade. In 1880 the college gave institutes on a more regular basis, but there never was the degree of central organization developed in Illinois. The Iowa expenses rested on the local groups until the legislature granted $50 to each county in 1891. The Iowa State Agricultural Society reports of 1898, 1899, and 1900 included accounts of institute meetings and papers read at particular ones.[25]

Through the latter half of the century, men in periodical offices, government agencies, and the land-grant colleges tried to help Illinois and Iowa cattle feeders improve their beef cattle, their types of feed, and their feeding practices. The intention was to help men raise beef in the most economical way by providing information about the many variable factors involved in the profitable feeding of beef cattle. Men sought for answers to many questions: which, if any, breed of cattle gained weight faster than others on the same feed, what were the comparative rates of gain and costs for various feeds, was ground corn better than whole kernel corn? Although complete and correct answers to these and many other questions were not always available, and information was not always disseminated in a way to be practically useful, men in the institutions of improvement probably were a source of help to Illinois and Iowa cattle feeders.

CHAPTER 10. AGE OF IMPROVEMENT III: CULMINATION

The feedlot was the testing ground for cattle improvement because all the publicity, experiments, and capital invested in producing early maturation would come to nought if the feed and feeding routine did not complement the other parts of the grand design. This was particularly important in the late 1880s and early 1890s when cattle prices declined. The need for rapid and economical gains in weight forced men to be more conscious of relative costs of different feed combinations.

In the decade before 1870, feeding practices continued much the same as in the 1850s. In many parts of Illinois and Iowa, men still fed mostly grass, and all depended heavily upon summer pasture even when corn was the major feed.[1] Aside from grazing, feeding practices varied as some turned cattle into standing corn (though this seemed to be on the decline), while others cut the corn and fed it as fodder, ear and stalk together. Some feeders raised their own corn and frequently bought husked or standing corn from neighbors. The majority, however, seemed to adhere to "stalk feeding" in which, after grass failed in the fall, cattle were concentrated in small fields or lots and fed ear corn, stalks, and hay thrown on the ground. After a suitable time swine were run into the lots to scavenge after the cattle.[2] The basic plan had variations as men experimented. Morris Case, near Waterloo, Iowa, had 150 head of cattle in rows of stalls facing a track used to distribute feed ground by a steam engine. Others built troughs or feed bunks 3 feet off the ground and 10 to 14 feet long for snapped ear corn and hay, and a few advocated use of oil cake meal, bran, oats, and even turnips as supplements.[3]

There was still much argument over whether animals should be sheltered, a favorite journalistic subject in the 1840s and 1850s. Some writers still contended that feeding cattle in open lots was "radically wrong, inhuman, and un-profitable,"[4] but despite the preaching and the supposed economic advantages, very few men kept their cattle in barns. As one farmer put it, "where corn and cattle are both easily produced, we must adopt a plan by which to feed our corn with as little handling as possible."[5] The usual shelter was a grove of trees on the north and west sides of the feedlots, a board fence, or a rude pole shed with straw piled on it.

Summer grazing continued on undrained prairie and marsh areas. In 1871 a Lee County, Illinois, man reported the "farmers all turning their attention to corn, hogs, and cattle" and that during the summers over 8,000 cattle fed on the Winnebago "swamp." This free forage, he estimated, added about 300 pounds per head to three-year-old steers. In the 1870s, herd grounds still existed in eastern Iowa; in Kankakee and Iroquois counties, Illinois; and in adjoining Lake, Newton, and Benton counties, Indiana, but even these diminished in size each year[6] because of increasing settlement. Some range herding continued in Illinois into the 1880s, but the main range country was western Iowa, where men herded cattle for a season at $2 a head, though even that acreage shrank continually. By 1873 there was little open prairie east of the Des Moines River, and by 1880 only a very few herds grazed on prairie in the three tiers of counties in southwest Iowa. In the northwest men grazed cattle on open range until 1885, but railroads brought that to an end by opening the region for raising grain.[7]

With the loss of the range or prairie option, feeders turned to improving their fenced pastures or feeding more corn.[8] For those who kept grass as an important option in the feed regime, bluegrass and timothy replaced common prairie grass in the 1870s and later. Even so, a late spring or summer drought could have serious consequences for a man depending on pasture. Men often kept more cattle during the winter than they could full-feed on stored corn and hay beyond the usual beginning of spring, and a dry summer would reduce the carrying capacity of pasture quickly, particularly of common prairie. In a dry year, half-fattened stock had to be sold early in the fall, or cattle turned into cornfields in August.[9]

Bluegrass pasture in some cases also provided winter grazing. In the 1870s, one prominent central Illinois feeder who used bluegrass as the major feed for his fat cattle argued (but did not prove) that 100 acres of bluegrass could equal 66 acres of corn in fattening value. In the middle 1870s James N. Brown's sons sold 700 to 900 steers annually from their 3,000 acres of bluegrass. They bought feeder cattle in the fall and kept them 9 to 12 months for a gain of 350 to 400 pounds. They claimed that on only about 11 days during the year did they have to put out hay because snow had covered the grass.

The Browns fed Texas cattle from 1866 to 1873, but after that went back to grade Shorthorn steers bought at two and three years of age at 1,000 pounds. They sold their cattle at three and four years and cleared about $8 a head on the operation. But the Browns soon turned to preparing cattle for market at a younger age. In 1878 at the first Fat Stock Show, the Browns won first and second prizes with steers 21 and 22 months old that showed a gain of 2.28 and 1.9 pounds per day. The judges declared the first prize winner a "superior" beef animal with flesh "much riper than the average three and four-year old steers in our markets."[10]

The Browns were exceptions in depending on bluegrass for fattening cattle because most men fed corn to steers for at least several months in the late fall and winter. Many still turned the cattle loose in the cornfields to forage for themselves, especially if there was a light corn crop and it did not seem to pay to husk or shock it. Some Iowa men custom-fed cattle by renting out their cornfields with the fee based on the weight added to the cattle.[11]

The head farmer at the Illinois Industrial University insisted that the best way to feed beef cattle was to stable them and feed them cornmeal mixed with straw and hay, but few appeared to follow his advice in practice. All agreed, nevertheless, that corn was the grain to feed; and editors urged men to "never as a general rule, raise corn to sell, but feed it to good, thrifty stock of some kind."[12] When the price of corn fell, it paid farmers to feed it rather than sell it. Because corn more often than not was a surer crop than wheat, farmers produced more and more corn, which created an increasing demand for feeder cattle and hogs. During 1873 the Prairie Farmer market writer warned farmers not to ship half-fattened cattle to market. Because of cheap corn, he warned that there would be more cattle in the early spring market than usual, and as a result prices would decline for half-fattened stock faster than for fully fattened cattle.[13] The crop was so good one man even protested that it was "no use to tell farmers to feed the corn crop," there was just too much. The abundance even encouraged men to feed Texas cattle, in spite of the tick fever and their relative slowness to take on fat in the winter.[14]

While most Illinois feeders in the middle 1870s finished their beef on corn, some Iowans depended more on grass. Some men grazed cattle for two years on prairie or pasture and did not put them into the feedlot until the fall of the third or even fourth year. There they ate corn and hay for several months and were marketed when prices and finish were right for the individual feeder.[15]

The question of winter shelter for cattle continued to be ignored in the 1870s. Few farmers reported feeding cattle in a barn despite illustrated articles on barn building in the press or claims that it was cheaper in the long run. Northern Illinois and Iowa counties reported hard winters, which retarded the fattening process and caused as much as a two-month delay in spring marketing. The quality of winter feed, especially corn, determined how fast cattle gained weight and when they were ready for spring markets. Soft corn or a shortage affected the quality of the beef in the spring. In the case of a short crop many men fed what corn they had early in the fall and sent their beef and pork to market before winter to avoid purchasing grain at high prices.[16]

In the 1880s there were some changes in feeding practices as cattle raising declined, especially in Iowa, because farmers drained prairies and planted corn. Cattle fattening, however, increased as more men fed cattle not calved on their farms.[17] Illinois farmers reported feed practices ranging from use of grass alone to mixtures of ground corn and oats or additions of bran. Ear, shelled, or ground corn now prevailed over the old "stall" feeding method. A Henry County, Illinois, man reported a movable self-feeder holding a month's supply of shelled corn for a carload of cattle. Many men favored the third year for marketing, but some had begun to sell beef at 21 to 24 months of age at weights up to 1,600 pounds.[18] Although some Iowans still sold four-year-old beef, on the whole, practices did not differ from the changes taking place in Illinois, with the exception of more grazing in western Iowa. There in 1880, W. H. Fitch of Calhoun County marketed two carloads of grade Shorthorn steers that had been grazed on prairie during the summers.[19]

There was increasing evidence that feeders realized the

advantages of genetic improvement and feed changes for profitable marketing of cattle at a younger age. The greater capital investment of this system was balanced by the faster turnover in the capital. Tom C. Ponting (Illinois feeder and Hereford raiser) fed his prize 21-month-old steers for the sixth Fat Stock Show on pasture, hay, shelled corn, oats, and oil cake the last nine months.[20] John M. Stahl (a central Illinois feeder) argued that the increased cost of maintaining an improved steer made it impossible to first grow steers and then fatten them; they had to be fattened from birth. The best way to get quick maturity was to start feeding oats, bran, and pasture before weaning; good summer pasture the first year; then corn the spring of the second year. This routine would produce 1,500-pound beef steers by 24 months.[21] Of the six yearling entries in the cost-of-production ring at the sixth Fat Stock Show, in the first year all had been fed on milk, hay, and pasture; four on shelled corn also; three on oats; and one on oil meal. In the second year the standard formula was shelled corn, hay, and pasture; five had oats in addition; and one of those also had oil cake. Of the four steers over two and under three years, one had only shelled corn, hay, and pasture each year; two had oats, oil cake, and oil meal in addition; and the fourth had bran and shorts but no oats in addition to corn, hay, and pasture.[22] John D. Gillett said in an interview in 1883 that it did not pay to produce beef over three years old. He planned to wean calves on oats and add a corn ration before 12 months in order to obtain weights of 1,500 pounds between 20 to 28 months of age.[23]

University of Illinois experimenters concluded in 1885 that young animals made the best gains and that a combination of grazing and a finishing period of three months on grain made the best gains and profits. The National Livestock Journal answered a mailed question by saying that a ration consisting of a slightly dampened mixture of five pounds of cornmeal, five pounds of bran, and two pounds of linseed meal mixed with two bushels of cut straw or corn fodder would finish 1,000-pound steers for the fall market in about three months.[24]

By the 1890s "albuminoids, carbohydrates, [and] the nutritive ration" had replaced fancy bloodlines as the topic of discussion according to the Iowa Homestead. Cattle feeders were more aware of the properties of different foods as cattle prices remained low and feeders became more economy conscious. They learned more about balancing supplementary foods such as oil meal, crushed oil cake, bran, and molasses with corn and grass, although there was some danger of spending money needlessly on "condimental preparations for cattle" which were not basic foods according to _Wallaces' Farmer_. Feeders began to use clover and alfalfa and were urged to use chopped corn as economical fodder.[25]

By 1900 agricultural experiment station reports had produced some statistical results of feeding experiments which could help cattlemen even if only as confirmation of existing practices. In 1898 W. A. Henry (dean of the University of Wisconsin College of Agriculture) published a compilation of the results of feed experiments in the previous decade. (_Wallaces' Farmer_ urged readers to order Henry's book.)Henry concluded that whole corn fed in the husk was the best feed when corn was cheap. Cornmeal made better gains on steers per day but was not good for fattening hogs that followed cattle, and in addition, the cost of grinding the meal added to the overhead. Oil meal was more expensive but gave better balance to the ration by adding protein to the carbohydrates in the corn. Grain fed in addition to good lush pasture did not appear to pay, but by using a grain supplement, the pasture needed for summer could be reduced by one-half.[26] By the end of the century, corn was still the major cattle feed and the feeder had to balance costs of feeder cattle, pasture, grain, and supplements with the rapidity of gain and the price of finished beef to decide which combination of feeds to use. A ration of cornmeal, oil meal, bran, and shorts with corn fodder required 28% less grain for 100 pounds of gain but cost 28% more than an ear-corn ration in the middle 1890s. The feeder had to decide which was the more economical way to fatten his beef in the given circumstances. Cattle made faster gains if fed cornmeal during the whole fattening period, but hogs in the feedlot could not utilize corn in the form of meal. Farmers had to balance the value of the better gains from cornmeal against the loss of gains in swine. Bran gave the feed ration bulk and protein, and when added to either whole or ground corn, it produced better gains than corn alone. Oil cake made from flaxseed was also an excellent feed supplement. Molasses was considered a good supplement, but it was not widely used. Locally grown clover and alfalfa also were coming into general use for fodder. These two legumes provided good protein and carbohydrate additions when fed as hay. In all cases of feed supplements,

120

cattlemen had to balance the increased gains in weight against the increased cost of the feed. Depending upon prices for bran, oil cake, corn, and feeder cattle compared to beef prices, cattlemen could choose the type of ration to produce the most economical gains in weight.[27]

Indeed, as Henry concluded, "the ability to fatten cattle rapidly and profitably is a gift," and feeders who had it were able to balance alternatives, suit the food to the condition of the cattle, bring cattle onto full feed, and keep them there with a carefully established routine. By shortening the time available for producing prime beef, the margins for error shrank. Anything that disturbed cattle once they were started on the feed routine would reduce profits or even cause a loss. Dean Henry cautioned men to be regular in their feeding program. They should not change the time of feeding, the place of feeding, or the people doing the feeding, or it would put the cattle "off feed." The ability to take cattle from the range, place them in a lot, and then quickly bring them to full feed and keep them there for three to six months required management skill; the unskilled fell by the way.[28] A feeder with a "good eye" could tell by looking at his cattle if they were gaining as they should and what should be done to maintain this.

Good feeders understood the place of oats, oil meal, bran, and corn fodder in feeding rations; the advantages of different types of corn preparation; the fact that cattle gained less at a higher cost as they increased in age; and the importance of adjusting these combinations as the cost of feeds and price of beef changed.[29] They combined skill in feeding with cattle bred for early maturation to produce quality beef at an age their grandfathers would have thought impossible. By the end of 1900 the cattle feeding business had changed so much that the heavy cattle for which Gillett was famous would no longer bring awe and praise in the market. The "beautiful" heavy steers of the past had lost favor and indeed suffered the additional blow of price discrimination. Buyers did not want steers much over 1,600 pounds liveweight. As a leading commission firm at the stockyards reported, "the heavy cuts are out of style and all classes of buyers would prefer lighter beeves."[30] The improvers had succeeded in changing the industry.

121

CHAPTER 11. ILLINOIS-IOWA CATTLE FEEDING IN THE NINETEENTH CENTURY

The development of the cattle feeding business detailed here occurred during the time when the United States went from an agricultural to an industrial economy and thus must be thought of in that larger context. Major changes in the transportation system and in population distribution and density created new business opportunities for both industry and agriculture that encouraged changes in production and organization. As industrial growth gave rise to production, distribution, and administrative change that encouraged corporations and "trusts," the Illinois and Iowa cattle feeding business also changed. In the immediate sense it grew up and changed with the Corn Belt, but it had also grown up with the national economy. The railroad-created national market and the growing urban complex demanded goods and services on an unprecedented scale, which encouraged and afforded the opportunity for change within agriculture.

The rich agricultural resources of Illinois and Iowa for producing corn and pasture made it an excellent area for beef cattle production. Settled by people familiar with cattle feeding operations and well positioned in relation to population movement and the expanding railroad network, Illinois and Iowa became the center of prime beef production in the nation by the end of the century.

As early as 1850 men had established a pattern of cattle production brought from Ohio and Kentucky. Cattlemen fed corn and grass to four- to six-year-old cattle, which were either driven to eastern markets or sent to some midwestern packer to be slaughtered, pickled, and shipped to retail markets in barrels.

In Ohio there had been a progression from range to feeding area and then a decline in the importance of corn fattening of cattle because of its rise in Illinois and Iowa. In the latter two states, however, there was little decline in fat cattle production as the industry spread to the Great Plains. The center of fat cattle production did not move westward again. Instead there was regional specialization as ranchers produced young cattle for the feedlots of Illinois and Iowa. The fact that these two states continued as the center of prime beef production was indicative of one of the significant changes in the development of the Corn Belt economy during the nineteenth century. Cattle feeders shifted their operations in response to technological and market changes during the course of the century so that the center of prime beef production remained in the same area from 1860 to well into the twentieth century.

Railroads (by centralizing markets and providing the nucleus for the great development of the beef processing industry in Chicago) and the development of the refrigerator car made possible the growth of the dressed beef trade and brought changes to the pattern of beef cattle feeding. Before the railroads and refrigeration, the market for Illinois and Iowa beef was limited to a slow expansion. Barreled beef was processed only during the cold months of the year and was not as appealing to consumers as fresh beef. The fresh beef trade was in the hands of the local butchers, who slaughtered cattle only as there was demand because of the problems of preserving fresh meat. Butchers obtained their beef from large stockyards or bought it directly from farmers. They took the whole animal and made up for the waste material such as hide, tallow, and bone, which they could not use, in the price of the meat offered for sale.

The expansion of the railroads, the rise of the urban market, and the development of techniques of refrigeration changed all this. It became possible to ship fresh meat raised in Illinois and Iowa and slaughtered at any time of year in Chicago to any place reached by a railroad. This made it possible and desirable to develop a central slaughter market in order to realize benefits in purchasing cattle on a large scale and utilizing the large volume of former waste material in the manufacture of by-products. At the same time that the Chicago packers developed a central cattle market and processing center, they also reorganized the wholesale distribution of fresh meat. The ability of the packer to quickly and cheaply supply almost any part of the country and the foreign market with fresh beef led to greater consumption of beef and increased demand for fat cattle.

In response to this change in market and increased demand, both historical and geographical changes took place in the cattle feeding industry. During the rest of the century Illinois and Iowa men improved the cattle they raised or bought improved range cattle to fatten. They also changed their feeding practices in an effort to produce better beef in a shorter time. With competition from more cheaply raised and grazed range cattle, Illinois and Iowa men concentrated on producing, in a short time, corn-fed cattle for the top beef grades in the market.

An important factor in this change in feeding practices was the close connection between cattle feeders and the institutions of improvement. Cattlemen participated in newspaper discussions of changes in feed routines, were in the forefront in the formation of associations for improved farming in general and livestock production in particular, and had close connections with state agricultural societies, fairs, agricultural schools, and ultimately with the agricultural experiment stations. Leaders in the movement toward breed improvement and early maturity in beef cattle contributed their knowledge to the farm community and benefited from the technical studies of feeds and feeding that became available toward the end of the century.

Within the two states there were also shifts in the areas of concentrations of beef cattle. On a county basis, the center moved from central Illinois to northwestern Illinois and east-central and southwestern Iowa by the end of the century. However, the geographical expansion of the cattle industry continued. Men saw the pattern of operations in Illinois and Iowa and modified it for use in eastern Kansas and Nebraska or introduced new factors. The major new element was the spread of cattle across western range country. The impact of this expansion in area and numbers of cattle influenced changes in cattle feeding in Illinois and Iowa toward the end of the century.

By the 1890s the midwestern cattle feeding industry had reached another period of stability in its organization and operation. The three-stage process of moving cattle from range through feedlots to market was well defined. Range areas specialized in producing and raising young cattle. Illinois and Iowa farmers specialized in fattening young range cattle on corn, alfalfa, oil meal, bran, and other feed supplements. Compared to the 1850s cattle were marketed at a younger age, and when slaughtered, they produced a higher percentage of beef to total weight. The animals themselves were different because of improvements in breeding. They made greater gains in weight for a given amount of feed and carried a better distribution of flesh on the carcass (which gave more of the better cuts of beef) than did cattle marketed in the 1850s.

This is not to say that the transition was a smoothly accomplished feat. That agriculture in general and cattle feeding in particular benefited from some of the changes in the national economy is not to be denied, but they were also victims. Farmers found themselves competing in a world market where supply and demand relationships changed quickly and the output of many agricultural units caused a price drop. But farmers could not respond as did some industrialists and organize "trusts" to control competition or to increase efficiency. Many farmers suffered through periods of low prices and incomes after 1870 when competition from Australia and Argentina and from lower ocean freight rates caused problems for American producers in a world market; and railroad development and the rise of a national market and relatively low capital requirements encouraged easy entry of domestic competition, especially in grain production.

Agriculture also suffered during the depression years of the mid-1870s and early 1890s. Within the broader context of American agricultural development, Illinois and Iowa cattle feeders were a small unit but shared many of the problems and benefits as well as having some special rewards. Cattle feeding was a high-risk enterprise with limits imposed by managerial talent and financial resources. But cattle feeders had some comfort in knowing that their numbers were limited by higher capital requirements than general farming and by the skill requisite to handling livestock successfully. Although unable to profitably combine into large-scale producing units, cattle feeders did not suffer as much as grain farmers in the plains from inability to create a producer "trust" in competition with the various processor "trusts" of the time. Industrial development, urban growth, and the railroad network all worked to raise demand for quality dressed beef. While generally at the mercy of the transport system, on occasion, cattlemen benefited from the competition of the industrial giants, for example, railroad rate wars in 1877, 1881, 1884, 1887, and 1888.

Within this context at the end of the century, the cattle feeders of Illinois and Iowa as a group found themselves in a more complex market that required management ability, capital, and some luck; but the market structures, cattle

types, and feed routines were well defined. By the 1890s
Chicago shared the market with Kansas City, Omaha, and
smaller regional markets; but this did not change the system
so much as to benefit the producer in western Iowa. The success of the grain-livestock combination in the Middle West
may even have given the region a relative prosperity in the
hard times often associated with farming in the 1880s and
early 1890s; and when economic conditions began improving by
1896, the cattle feeders were in a good position to benefit.
The efficient feeders survived the period of declining profit
margins in the 1880s and early 1890s and established the
dominant pattern of cattle feeding operations.

The fattening of prime cattle was an art as well as a
science, and from 1840 to 1900 Illinois and Iowa cattle feeders had brought this to a state of performance and a pattern
of operation that in many ways remained the norm for the industry until the 1950s. For indeed, cattle feeding did not
decline in Illinois and Iowa after 1900. As late as 1959,
cattle and calves accounted for 28.1% and 42.2% of the value
of all farm products sold in Illinois and Iowa respectively,
while hogs accounted for only 18.1% and 24.4% of the value
of farm products sold in Illinois and Iowa.[1]

These figures and the foregoing discussion for the nineteenth century might well indicate a needed revision in the
standard terminology of "corn-hog belt." There is no denying
that hogs were important contributors to farm income in the
nineteenth century, but in Illinois and Iowa hogs often were
an adjunct to cattle feeding. Feeding cornmeal was never
very popular because meal could not be utilized by hogs as
whole corn could, and cattle feeders were quick to point
this out. From the 1830s corn and cattle were a natural and
rewarding combination for Illinois and Iowa farmers, and less
experimenting was required in these two states than in others
close by for farmers to find a commercial base for their
agriculture. The ingredients for the corn-cattle belt were
present from the early part of the nineteenth century, and
farmers used them to continue cattle feeding as a major occupation to the present day by changing their operations to
meet the changing market needs. Rather than being a passing
phase of frontier agricultural development as earlier in
areas to the east, cattle feeding in the Corn Belt was a
commercial venture very early and remained a major element
in the economy of Illinois and Iowa in the middle of the
twentieth century.

A P P E N D I X A. MAPS

Map 1. Iowa and Illinois Counties

Map 2. Area of Wisconsin Glaciation

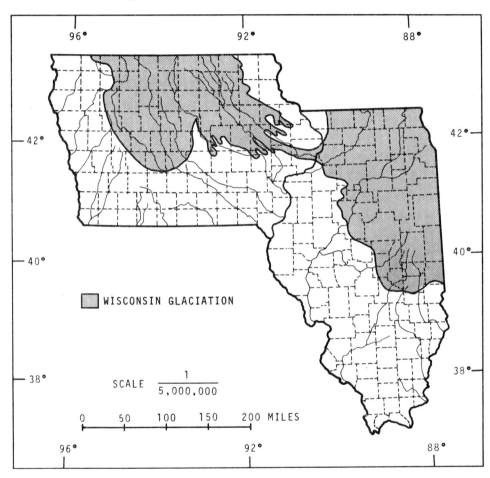

Map 3. Illinois-Iowa General Soil Types

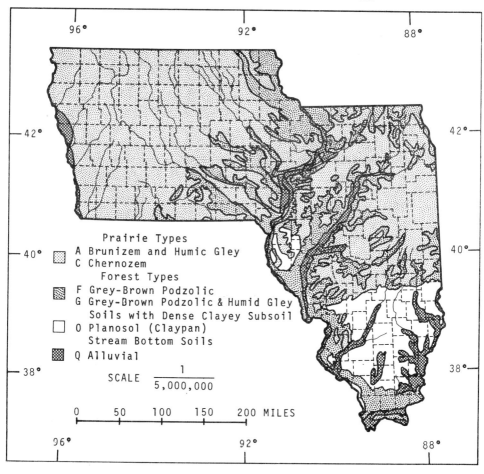

Prairie Types
A Brunizem and Humic Gley
C Chernozem
    Forest Types
F Grey-Brown Podzolic
G Grey-Brown Podzolic & Humid Gley
    Soils with Dense Clayey Subsoil
O Planosol (Claypan)
    Stream Bottom Soils
Q Alluvial

SCALE $\dfrac{1}{5,000,000}$

0   50   100   150   200 MILES

Map 4. Illinois-Iowa Land in Farms

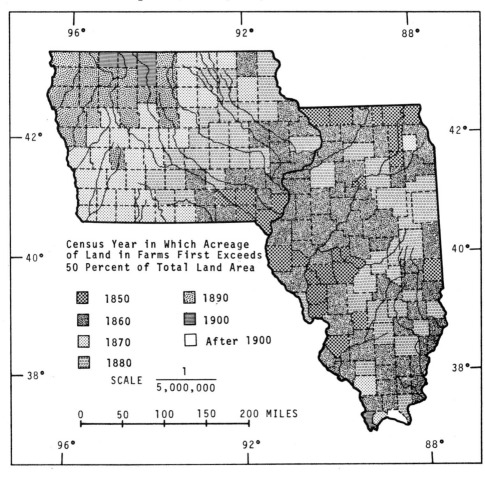

Census Year in Which Acreage
of Land in Farms First Exceeds
50 Percent of Total Land Area

1850   1890
1860   1900
1870   After 1900
1880

SCALE   1 / 5,000,000

0   50   100   150   200 MILES

Map 5. Railroads, 1861

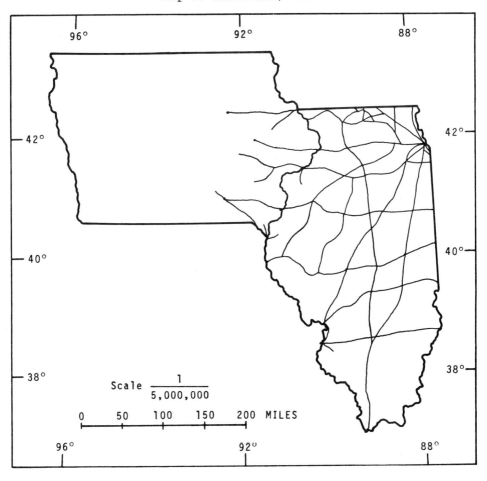

Scale $\dfrac{1}{5,000,000}$

131

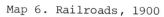

Map 6. Railroads, 1900

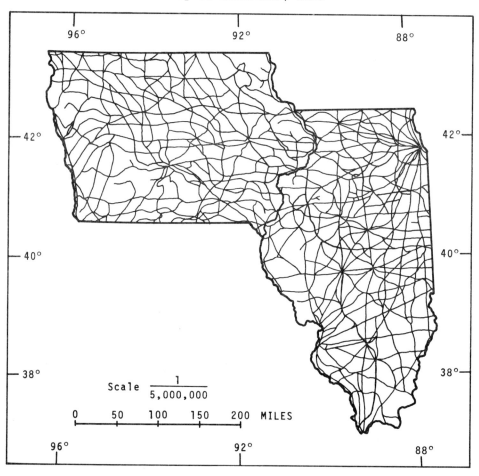

Scale $\dfrac{1}{5,000,000}$

0   50   100   150   200   MILES

Map 7. Beef Cattle per Hundred Acres, 1850

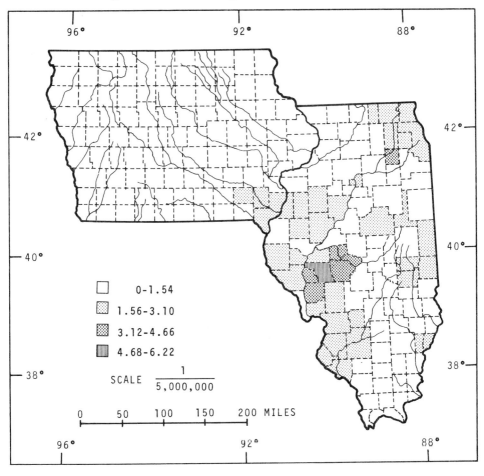

0-1.54

1.56-3.10

3.12-4.66

4.68-6.22

SCALE $\frac{1}{5,000,000}$

0   50   100   150   200 MILES

Map 8. Beef Cattle per Hundred Acres, 1860

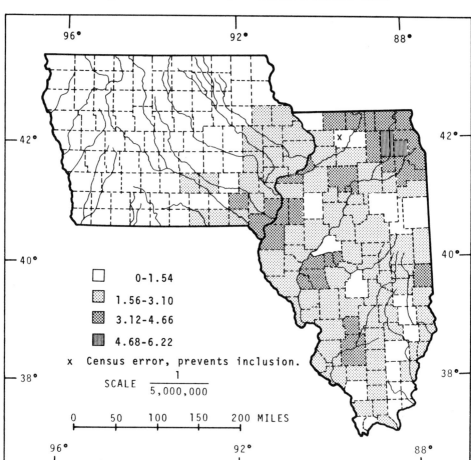

0-1.54

1.56-3.10

3.12-4.66

4.68-6.22

x   Census error, prevents inclusion.

SCALE $\dfrac{1}{5,000,000}$

0   50   100   150   200 MILES

Map 9. Beef Cattle per Hundred Acres, 1870

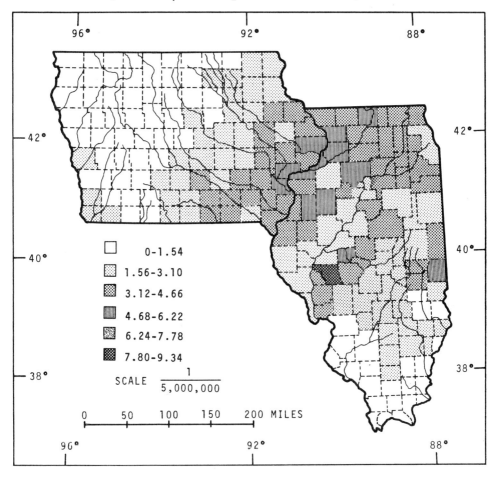

0-1.54
1.56-3.10
3.12-4.66
4.68-6.22
6.24-7.78
7.80-9.34

SCALE $\dfrac{1}{5,000,000}$

0   50   100   150   200 MILES

Map 10. Beef Cattle per Hundred Acres, 1880

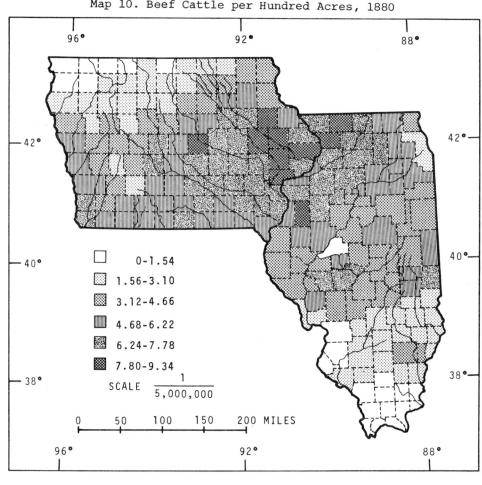

0-1.54
1.56-3.10
3.12-4.66
4.68-6.22
6.24-7.78
7.80-9.34

SCALE $\dfrac{1}{5,000,000}$

0   50   100   150   200 MILES

Map 11. Beef Cattle per Hundred Acres, 1890

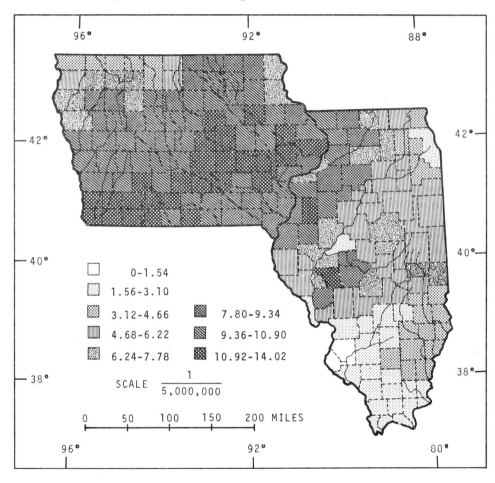

☐ 0-1.54	
1.56-3.10	
3.12-4.66	7.80-9.34
4.68-6.22	9.36-10.90
6.24-7.78	10.92-14.02

SCALE $\dfrac{1}{5,000,000}$

0   50   100   150   200 MILES

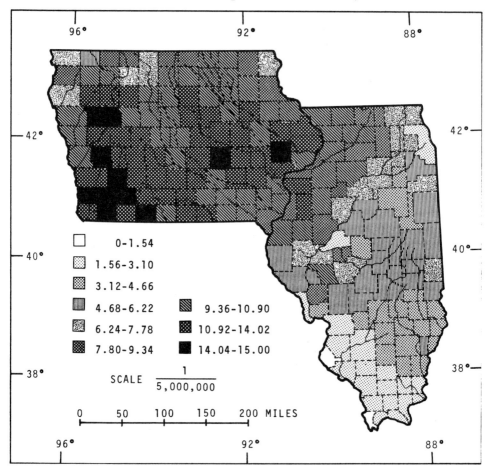

Map 12. Beef Cattle per Hundred Acres, 1900

□	0-1.54
	1.56-3.10
	3.12-4.66
	4.68-6.22
	6.24-7.78
	7.80-9.34

	9.36-10.90
	10.92-14.02
	14.04-15.00

SCALE $\dfrac{1}{5,000,000}$

0   50   100   150   200 MILES

Map 13. Registered Cattle Calved or Imported, 1870 (Source: Brinkman, "Historical Geography," 1964; map 28, used by permission)

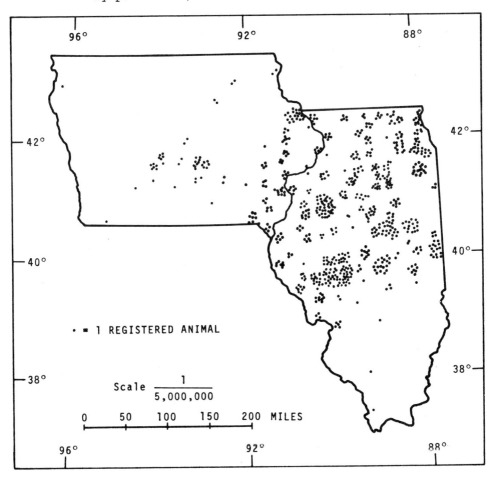

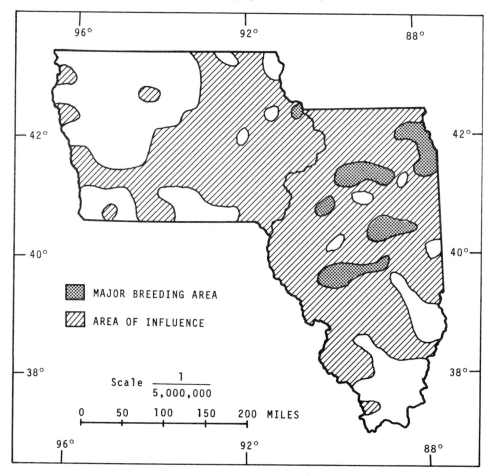

Map 14. Presumed Areas Influenced by Improved Breeds of Cattle, 1870 (Source: Brinkman, "Historical Geography," 1964; map 26, used by permission)

MAJOR BREEDING AREA

AREA OF INFLUENCE

Scale $\frac{1}{5,000,000}$

0   50   100   150   200   MILES

A P P E N D I X B. TABLES AND FIGURES

TABLE B.1. POPULATION GROWTH

State	1840	1860	1900
Illinois	476,183	1,711,951	4,821,550
Iowa	43,112	674,913	2,231,853

Source: Twelfth Census, Abstract, pp. 32-33.

TABLE B.2. ILLINOIS-IOWA POSITION AS A PERCENTAGE OF TOTAL U.S. AMOUN 1900

Land area	Popu- lation	Improved farm- land	Wheat harvested	Neat* cattle on farms	Swine	Corn harvested	Oats harveste
3.75	9.15	13.8	6.5	11.3	24.7	29.3	36.9

Source: Twelfth Census, Abstract, pp. 32-33, 234, 238, 245, 258-59.

*Neat = common domestic bovine.

141

TABLE B.3. ILLINOIS-IOWA AGRICULTURE: NATIONAL POSITION, 1850

Item	Illinois Amount	Per-cent, U.S. total	Iowa Amount	Per-cent, U.S. total
Number of farms	76,208	5.26	14,805	1.02
Improved land in farms	5,039,545 acres	4.46	824,682 acres	0.73
Corn	57,646,984 bu	9.7	8,656,799 bu	1.5
Wheat	9,414,575 bu	9.4	1,530,581 bu	1.5
Oats	10,687,241 bu	6.9	1,524,345 bu	1.0
Swine	1,915,907	6.3	323,247	1.06
"Other" cattle	617,365	5.4	90,917	0.78

Source: Twelfth Census, V, pp. 704-5, 707 and VI, pp. 80-81, 84-85, 92-93; Thirteenth Census, V, pp. 69, 73-74 (crop year 1849; farms, land, and livestock of 1850).

TABLE B.4. ILLINOIS-IOWA AGRICULTURE: NATIONAL POSITION, 1870

Item	Illinois (% of U.S. total)	Iowa (% of U.S. total)
Number of farms	7.6	4.3
Improved land in farms	10.2	4.9
Corn produced	17.1	9.1
Wheat produced	10.5	10.2
Oats produced	15.2	7.4

Source: Thirteenth Census, V, pp. 69, 73-74; Twelfth Census, VI, pp. 80-81, 84-85, 92-93.

TABLE B.5. ILLINOIS-IOWA AGRICULTURE: NATIONAL POSITION, 1900

Item	Illinois (% of U.S. total)	Iowa (% of U.S. total)
Number of farms	4.6	3.9
Improved land in farms	6.6	7.2
Corn produced	14.9	14.4
Wheat produced	3.0	3.5
Oats produced	19.1	17.8
"Other" cattle on farms	3.8	7.5
Swine on farms	9.4	16.8

Source: Twelfth Census, V, pp. 704-5, 707 and VI, pp. 80-81, 84-85, 92-93; Thirteenth Census, V, pp. 69, 73-74.

TABLE B.6. SELECTED CHICAGO BEEF PRICES, 1840-60

Date	Price per cwt	Date	Price per cwt
1840 April	$4.00-4.50	1852 January	$4.00-4.50
July	4.00-4.50	April	4.00-4.50
1841 January	4.00-4.50	July	4.00-4.50
April	3.00-4.00	October	4.00-4.25
July	3.00-3.50	1853 January	2.50-4.00
1844 April	3.00	April	3.00-3.50
July	3.00	July	2.75
1845 January	2.50-2.62	October	2.50-2.75
April	3.50-4.00	1854 January	3.00-3.12
July	2.50-3.00	April	3.75-4.50
October	2.00-2.50	July	3.50-3.62
1846 January	2.00-2.50	October	2.75-3.12
April	2.50-3.50	1855 January	3.00-3.25
July	2.50-3.50	April	3.50-4.00
October	2.37-2.50	July	3.50-4.00
1847 January	2.25-3.00	October	2.50-3.12
April	3.50-4.00	1856 January	3.00-4.50
July	3.50-4.50	April	4.25-4.50
October	3.00-3.50	July	3.00-3.75
1848 January	2.25-3.25	October	2.00-2.50
April	2.50-3.50	1857 January	3.00-3.75
July	3.50-4.00	April	3.87-4.25
October	2.50-3.50	July	4.25-4.65
1849 January	2.50-3.50	October	2.50-3.00
April	3.00-4.00	1858 January	2.25-2.75
July	3.00-4.00	April	2.75-3.25
October	3.00-3.50	July	2.00-3.50
1850 January	2.50-3.25	October	1.50-2.50
April	2.50-3.25	1859 January	2.25-2.75
July	4.00-4.50	April	3.00-3.50
October	3.25-3.50	July	2.00-2.50
1851 January	3.00-3.50	October	1.50-1.75
April	3.00-3.50	1860 January	2.00-2.75
July	4.00-5.00	April	2.50-3.00
October	4.00-5.00	July	2.50-3.00
		October	2.25-2.50

Source: Compiled from Aldrich for 1840-44, 1854-57; Prairie Farmer, for 1845-53; Chicago Board of Trade, Annual Report, for 1858-60. It is assumed these are prices for liveweights.

Note: Prices are not fully comparable because there are no grade definitions in the sources. Some figures represent actual sales and others are averages. The ranges can generally be assumed to represent figures for average quality beef.

TABLE B.7. ILLINOIS-IOWA RAILROAD DEVELOPMENT: SQUARE
          MILES FOR EACH MILE OF TRACK

State	1860	1880	1900
Illinois	19.8	7	5
Iowa	84	10	5.8

Source: Calculated from Chicago Board of Trade, Report,
1900, pp. 41, 169 and 1901, p. 169.

TABLE B.8. RECEIPTS AND SHIPMENTS OF CATTLE AT CHICAGO UNION
         STOCK YARDS, 1865-1900

Year	Receipts (head)	Shipments (head)
1865	613 (5 days only)	0
1866	393,007	263,693
1867	329,188	203,580
1868	324,524	215,987
1869	403,102	294,717
1870	532,964	391,709
1871	543,050	401,927
1872	684,075	510,025
1873	761,428	574,181
1874	843,966	622,929
1875	920,843	696,534
1876	1,096,745	797,724
1877	1,033,151	703,402
1878	1,083,068	699,108
1879	1,215,732	726,903
1880	1,382,477	886,614
1881	1,498,550	938,713
1882	1,582,530	921,009
1883	1,878,944	966,758
1884	1,817,697	791,884
1885	1,905,518	744,093
1886	1,963,900	704,675
1887	2,382,008	791,483
1888	2,611,543	968,385
1889	3,023,281	1,258,971
1890	3,484,280	1,260,309
1891	3,250,359	1,066,264
1892	3,571,796	1,121,675
1893	3,133,406	900,163
1894	2,974,363	950,738
1895	2,588,558	785,092
1896	2,600,476	818,326
1897	2,554,924	843,392
1898	2,480,897	865,642
1899	2,514,446	811,874
1900	2,729,046	934,649

Source: Drovers' Journal, Year Book, 1932, pp. 10-11.

TABLE B.9. DRESSED BEEF SHIPMENTS FROM CHICAGO

Year	Dressed beef
	(tons)
1876	8,865
1880	15,680
1884	182,067
1888	384,741
1890	482,067
1895	455,169
1900	589,475

Source: USDA, Report, 1876, p. 320; Nimmo, Report on the Internal Commerce of the United States, Exec. Doc. 7, pp. 157-58; Chicago Board of Trade, Report, 1886-1900.

TABLE B.10. LIVE CATTLE AND DRESSED BEEF SHIPMENTS

Year	Live cattle		Dressed beef (Chicago)	Total dressed beef (Chicago)	Destination
	Chicago	St. Louis			
	(tons)				
1880				15,680	
	222,262	---	---	---	New York
	127,060	35,840	15,680	---	New England
1884				172,824	
	191,736	57,227	34,916	---	New York
	55,996	7,532	120,922	---	New England

Source: U.S. Congress, H. Exec. Doc. 267, pp. 155-58.

TABLE B.11. U.S. EXPORTS: FISCAL YEAR ENDING JUNE 30, 1900

Commodity	Value	Number
Total U.S. merchandise	$1,371,000,000	...
Raw cotton	242,000,000	...
Wheat and wheat flour	141,000,000	...
Total meats	114,000,000	...
Live cattle	30,623,768	396,977 head
Dressed beef	29,643,830	329,078,609 lb
Canned beef	5,233,982	55,553,745 lb
Pickled beef	2,893,902	49,622,328 lb
Total	68,395,482	

Source: Chicago Board of Trade, Report, 1900, pp. xxxvii-xxxviii, lvi; Bureau of the Census, Historical Statistics, pp. 537, 546.

Fig. B.1. Value of U.S. Exports of Pork, Lard, Live Cattle, and Beef, 1866-1900 by Five-Year Averages (Source: USDA, <u>Yearbook</u>, 1906, p. 248)

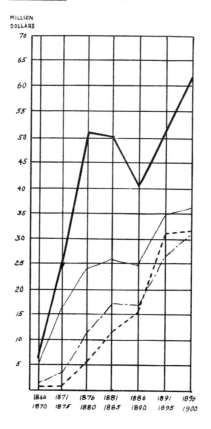

TABLE B.12. CHICAGO CATTLE RECEIPTS, SELECTED YEARS

Year	Texas and Cherokee	Western and northern range	Natives	Total	Range and Texas, % of total
1880	88,600	109,500	1,184,377	1,382,477	14
1882	346,300	229,700	1,006,530	1,582,530	36
1884	358,374	231,700	1,227,623	1,817,697	32
1886	320,830	238,520	1,404,550	1,963,900	28
1888	547,185	267,494	1,796,864	2,611,543	31
1890	657,053	229,494	2,597,733	3,484,280	25
1895	359,643	430,525	1,798,389	2,588,558	31
1900	194,726	147,000	2,387,320	2,729,046	13

Source: Drovers' Journal, Year Book, 1901, p. 22; ibid., 1932, pp. 10, 45, 46; Prairie Farmer, Jan. 4, 1890, p. 4.

Fig. B.2. Average Monthly Prices of Fat Cattle and Feeder
         Cattle, 1874-1900 (Source: Hopkins, Statistical
         Study, pp. 359-61)

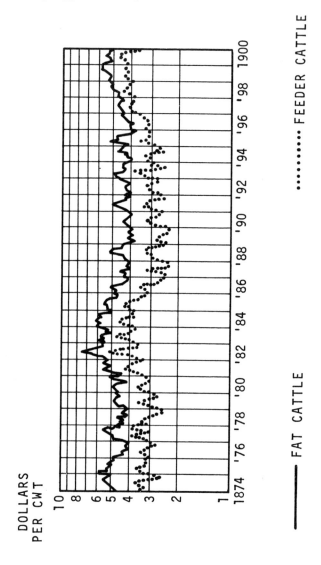

TABLE B.13. BEEF CATTLE ON FARMS

State	1850	1860	1870	1880	1890	1900
Illinois	617,365	1,061,179	1,075,265	1,518,409	1,975,233	1,373,024
Iowa	90,917	350,286	636,244	1,757,849	3,397,132	2,653,703
Percent of U.S. total "other" cattle						
Illinois	5.41	6.2	7.2	5.0	4.8	3.8
Iowa	0.79	2.0	4.2	6.4	8.2	7.5

Source: Twelfth Census, V, pp. 704-5.

NOTES

Preface

1. Thompson, <u>History of Livestock Raising</u>, USDA, Agr. Hist. Ser. 5, pp. 67-85, 99; Dunbar, Carolina Cowpens, <u>Agr. Hist.</u> 35(July 1961):125-31; Henlein, <u>Cattle Kingdom in the Ohio Valley</u>, pp. 1-10, 103-5, 110-13; Shannon, <u>Farmer's Last Frontier</u>, pp. 198-99.

Chapter 1

1. Wallace and Bressman, <u>Corn and Corn Growing</u>, pp. 89-92.

2. Garland, <u>North American Midwest</u>, pp. 6-11.

3. Odell et al., Soils of the North Central Region, Univ. Wis. Bull. 544, p. 7; Ross and Case, Types of Farming in Illinois, Univ. Ill. Bull. 601, pp. 7-9; for a more detailed discussion see Fenneman, <u>Physiography</u>, pp. 452-56, 472-82, 499-518, 559-60, 576-79, 594-605.

4. Odell, Soils of the North Central Region, pp. 6-9, 15-17, 25-26, 29-31, 36-37.

5. Kay et al., Pleistocene Geology of Iowa, Iowa Geol. Survey, pp. 11-15, 33-69; Odell, Soils of the North Central Region, pp. 6-7, 15-17, 25-26, 29-31, 36-37, 26.

6. Garland, <u>North American Midwest</u>, pp. 22, 25.

7. Nelson, "A History of Agriculture in Illinois"; Hawk, "Iowa Farming Types."

8. George Rogers Clark to George Mason, Nov. 19, 1779, in English, <u>Conquest of the Country Northwest of the River Ohio</u>, vol. 1, p. 453.

9. Shirreff, <u>Tour through North America</u>, pp. 451-52; Lea, <u>Notes on Wisconsin Territory</u> (Repr. State Hist. Soc. Iowa as <u>The Book That Gave to Iowa Its Name</u>, Iowa City, n.d.),

p. 28; Angle, Prairie State, passim; Harding, "Economic and Social Conditions in Iowa to 1880"; Bushnell, Work and Play, p. 236.

10. Twelfth Census, Statistical Atlas, 1903, pls. 4-8; Royce, Indian Land Cessions, H. Doc. 736, map 24; Pooley, Settlement of Illinois, Univ. Wis. Bull. 220, p. 274.

11. Pooley, Settlement of Illinois, p. 316; Royce, Indian Land Cessions, pp. 656-57, 664-67, 672-73, 678-81, 692-93, 696-97, 706-7, 722-27, 736-37, 750-51, 766-69, 778-79, 784-85, maps 17, 18, 24, 25; U.S. Commissioner of the General Land Office, Report, 1838, 25th Congress, 3rd sess., S. Doc. 17, map 10; idem, Report, 1843, 28th Congress, 1st sess., S. Doc. 15, p. 65; Lokken, Iowa Public Land Disposal, pp. 13-16, 26, 34, 45-46, 50-53.

12. Twelfth Census, Statistical Atlas, pp. 22-37, pls. 2-13; ibid., V, pp. 142, 322-37, 703-5, 707; ibid., VI, pp. 62, 68-69, 80-81, 84-85, 213.

13. Buckingham, Early Settlers, J. Ill. State Hist. Soc. 35:243; Goodwin, American Occupation, Iowa J. Hist. Polit. 17:89-92, 96-99; Census of Iowa for 1880, pp. 4-7; U.S., Statutes at Large, 2:58-59, 514-16, 743-47, 3:428-31, 536, 4:701, 5:10-16, 23-41, 9:117; Pooley, Settlement of Illinois, pp. 315-22; Thirteenth Census, 1910, V, pp. 73-74.

Chapter 2

1. Bogue, Prairie to Corn Belt, pp. 230-34; Danhof, Change in Agriculture, pp. 108, 121-25; Clark, Grain Trade, pp. 85-91, 100-101. Although Danhof made a major point that farmers had to change attitudes to become commercial while social pressure reinforced the attempt to be self-sufficient and thus inefficient (pp. 16-17, 19, 144), his comments about resisting specialization and slowness to be market oriented concern New England in the 1830s more than the midwestern prairies of 1850. When discussing the prairies, he acknowledged a quick shift to commercial agriculture (pp. 146-47, 150-52).

2. Lampard, Rise of the Dairy Industry in Wisconsin, pp. 47-57.

3. Springfield State Register, May 10, Aug. 2, 11, Nov. 15, 1849, Cole Notes; Anderson, "Agriculture in Illinois," pp. 148-49; Bogue, Prairie to Corn Belt, p. 234; Iowa Homestead, June 14, 21, Aug. 23, 1865; Prairie Farmer, May 15, p. 156, May 29, p. 176, Oct. 30, p. 360, Nov. 20, 1869, p. 384.

4. Cole, Era of the Civil War, pp. 82-83; Ill. State Agr. Soc. Trans., 1861-1864, pp. 317-82 passim in County Society Reports, hereafter this series will be cited as ISAST; Northwestern Farmer (Dubuque), Jan. to Aug. 1858, passim; Twelfth Census, VI, p. 477.

5. Cole, Era of the Civil War, p. 381; Prairie Farmer, Oct. 10, 1868, p. 116, Apr. 23, 1870, p. 124, Apr. 15, 1871, p. 116.

6. Schoolcraft, Travels in the Central Portions of the Mississippi Valley, pp. 303, 308, 310; Flint, Condensed Geography, vol. 2, pp. 120-21, 128; Shirreff, Tour through North America, pp. 445-47, 239, 242; Prairie Farmer, Feb. 1850, p. 45; Carman, English Views, Agr. Hist. 8(Jan. 1934): 3-19; Prairie Farmer, passim; Oliver, Eight Months in Illinois, pp. 102-4; the average size of farms as determined by the federal census became progressively smaller from 1840 to 1890.

7. Prairie Farmer, July 21, 1859, p. 41.

8. Anderson, "Agriculture in Illinois," pp. 310-14; Twelfth Census, V, pp. 704-8, VI, pp. 80-81, 84-85, 92-93; Thirteenth Census, V, pp. 73-74.

9. Bogue, Prairie to Corn Belt, pp. 218, 221, 226; Newlove, "Economic History of Illinois Agriculture," pp. 28-29; Clark, Grain Trade, pp. 85-86, 153-55.

10. Prairie Farmer, Sept., pp. 298-99, Nov. 1850, p. 334, Sept. 1, p. 132, Oct. 13, 1859, p. 233; Northwestern Farmer, Oct. 1858, p. 322; Anderson, "Agriculture in Illinois," pp. 77-83, 185-88; Clark, Grain Trade, pp. 202-5; ISAST, 1869-1870, pp. 142-43; Ill. Dept. Agr. Trans., 1875, pp. 297, 308-9, hereafter this series, which superseded ISAST, will be cited as IDAT.

11. Salesman quoted in Anderson, "Agriculture in Illinois," p. 123; Buck, Pioneer Letters of Flagg, Trans. Ill. State Hist. Soc., 1910, p. 182; Cole, Era of the Civil War, pp. 374-75; Mound City Emporium, May 13, 1858, p. 2, Cole Notes; Prairie Farmer, Aug. 18, 1859, p. 101, Dec. 5, 1868, p. 184, Oct. 30, 1869, p. 360; see Power, Planting Corn Belt Culture, for a discussion of the relationship between cultural attitudes and types of farming.

12. Cole, Era of the Civil War, p. 375; Prairie Farmer, May 7, 1870, p. 138.

13. Anderson, "Agriculture in Illinois," pp. 86-87; Cole, Era of the Civil War, pp. 39-47, 373-75; Taylor and Neu, American Railroad Network, map.

14. Taylor and Neu, American Railroad Network, map;

Mildred Throne, Population Study, Iowa J. Hist. 57:317-19, 321.

15. Bogue, Prairie to Corn Belt, pp. 219-22, 227; Hawk, "Iowa Farming Types," p. 20; for a more detailed discussion of agriculture in both states see also Anderson, "Agriculture in Illinois," Nelson, "A History of Agriculture in Illinois," Newlove, "Economic History of Illinois Agriculture," and Ross, Iowa Agriculture.

16. USDA, Yearbook, 1940, pp. 228-32; see any Prairie Farmer in the 1850s for machinery manufacturers' advertisements and claims; Prairie Farmer, Aug. 25, 1859, p. 116; Rogin, Introduction of Farm Machinery, p. 35, and pp. 31-52, 85-119, 133-45 for changes taking place in tillage and wheat harvesting implements; Shannon, Farmer's Last Frontier, pp. 125-72.

17. Putnam, Illinois and Michigan Canal, pp. 99-125, discusses the economic influence of the canal and the quickly developed railroad competition; Iowa Census for 1880, pp. 12-27; U.S. Interstate Commerce Commission, Thirteenth Annual Report, p. 12.

18. Ross, Iowa Agriculture, p. 49; Twelfth Census, V, p. 702; Thirteenth Census, V, p. 69.

19. Hawk, "Iowa Farming Types," pp. 63-77, 99-100, 106.

20. Nelson, "History of Agriculture in Illinois," pp. 96-102, 127-57.

21. Twelfth Census, Abstract, pp. 250-51, 258-59; ibid., V, pp. 322, 326-27; John King of northeastern Iowa in the Chillicothe Advertiser quoted in the Prairie Farmer, Sept. 1845, p. 215.

22. Twelfth Census, I, pp. lxxxiii, lxxxiv, lxxxviii.

Chapter 3

1. Thompson, History of Livestock Raising, USDA Agr. Hist. Ser. 5, pp. 67-85; Dunbar, Carolina Cowpens, Agr. Hist. 35:125-31; Shannon, Farmer's Last Frontier, pp. 198-99; Henlein, Cattle Kingdom in the Ohio Valley, pp. 1-10, 103-5, 110-13.

2. Schoolcraft, Travels, p. 310; Flint, Condensed Geography, vol. 1, pp. 82, 228-29, vol. 2, p. 128; Solomon Koepfli to J. A. Prukett, Apr. 13, 1861, Flagg Papers; Bateman and Selby, Historical Encyclopedia, p. 899; Shirreff,

Tour through North America, p. 446; see also Oliver, Eight Months in Illinois, and Massey, "British Travelers' Impressions of Agriculture," pp. 89-91.

3. Shirreff, Tour through North America, p. 249; U.S. Commissioner of the Patent Office, Report, 1849, pt. 2, p. 293, hereafter referred to as Patent Office, Report.

4. Henlein, Cattle Kingdom in the Ohio Valley, pp. 58-65, 116-20, 170; Harris, Autobiography, Trans. Ill. State Hist. Soc., 1923, pp. 72-101.

5. Patent Office, Report, 1852, pp. 340-41; A Pioneer, Northern Iowa, pp. 7-8; Henlein, Cattle Kingdom in the Ohio Valley, pp. 58-65, 116-21, 170; Bogue, Swamp Land Act, Agr. Hist. 25:169-80; Kerrick, Life and Character of Isaac Funk, Trans. Ill. State Hist. Soc., 1902, pp. 161-65; Hopkins, Economic History, pp. 68-69; Eighth Census, II, pp. cxxx-cxxxi.

6. Prairie Farmer, Sept. 1850, pp. 278-79.

7. Patent Office, Report, 1856, pp. x-xii; see also Prairie Farmer, July 1847, p. 201, July 1849, p. 216, Sept. 1850, p. 265, Dec. 1851, p. 576, Feb. 1852, p. 93, May 1854, p. 166; Northwestern Farmer, May 1858, pp. 166-67, July 1858, p. 230.

8. Prairie Farmer, market reports, May 22 to Aug. 7, 1856. This is based upon reports from New York City stockyards on the number and origin of cattle received each week.

9. Burlend and Burlend, True Picture of Emigration, pp. 75-78; [Ponting] Life, pp. 5-19; Kerrick, Life and Character of Isaac Funk, pp. 161-63; Eli Strawn to Joel Hopkins, Sept. 20, Oct. 17, 25, 1860, Hopkins Papers; Prairie Farmer, Nov. 1845, pp. 427-30; Bateman and Selby, Historical Encyclopedia, p. 965; Shaw Smith to Amos Williams, Feb. 28, 1834, Williams-Woodbury Papers; McCoy, Historic Sketches, p. 238; Chicago Daily Democratic Press, Oct. 25, 1854, quoted in Anderson, "Agriculture in Illinois," p. 146.

10. Northwestern Farmer, July 1858, p. 231.

11. Prairie Farmer, Aug. 1855, p. 248, June 25, Oct. 15, 1847; Cattleman 35:34-35; [Ponting] Life, pp. 28-34; New York Tribune, July 4, 1854, p. 8; Springfield Journal, July 17, 1854, p. 3, Cole Notes; Chicago Daily Democrat Press, Sept. 10, 1855, p. 2, Cole Notes; Henlein, Cattle Kingdom in the Ohio Valley, pp. 177-79; McCoy, Historic Sketches, pp. 38-41.

12. Prairie Farmer, Nov. 1849, p. 356.

13. Sanders, James N. Brown, Trans. Ill. State Hist.

Soc., 1912, pp. 172-73; Prairie Farmer, Jan. 3, 1856, p. 2; McClelland, Jacob Strawn, J. Ill. State Hist. Soc. 34:195-97; Ross, Iowa Agriculture, pp. 45-46; Northwestern Farmer, May 1858, pp. 166-67; Brinkman, "Historical Geography of Improved Cattle," maps 1-4, 8, 10, 12.

14. Prairie Farmer, Jan. 1843, p. 4, Oct. 1848, pp. 332-33, Sept. 1850, pp. 298-99, Nov. 1850, pp. 330, 334, July 1851, p. 317, Dec. 1851, pp. 561-64, Nov. 1854, p. 404, Feb. 12, 1857, p. 1, Jan. 27, 1859, p. 51, Dec. 15, 1859, pp. 370-71; Patent Office, Report, 1854, p. 6; ibid., 1855, pp. 20-21; ibid., 1858, p. 107; Northwestern Farmer, Mar. 1858, pp. 94-95, Nov. 1858, pp. 369-70; Illinois State Register, Apr. 14, 1853, p. 5, Cole Notes; ISAST, II:256-57, 276-77, 294-95, 332, III:133-34; Brinkman, "Historical Geography of Improved Cattle," map 8.

15. Thomas Le Duc, Public Policy, Private Investment, Agr. Hist. 37:8; Prairie Farmer, Apr. 19, 1857.

16. Kerrick, Life and Character of Isaac Funk, pp. 164-65; Cavanagh, Funk of Funk's Grove, pp. 49, 66-67.

17. Newcombe, Alson J. Streeter, J. Ill. State Hist. Soc. 38:422; Shaw Smith to Amos Williams, Feb. 28, 1834, Abram Stansbury to Amos Williams, Mar, 30, 1844, Williams-Woodbury Papers; Rammelkamp, Memoirs of John Henry, J. Ill. State Hist. Soc. 18:70.

18. Cavanagh, Funk of Funk's Grove, p. 66; McClelland, Jacob Strawn, p. 188; Harris, Autobiography, pp. 81-82; Eli Strawn to Joel W. Hopkins, Sept. 20, Oct. 17, 25, 1860, Hopkins Papers.

19. Prairie Farmer, Feb. 1843, pp. 45-46, Mar. 1843, pp. 50-51; Patent Office, Report, 1845, p. 1012; some of the feed recipes given sound like pudding for cattle, e.g., to 156 pounds boiling water add 2 pounds linseed meal, boil five minutes, add 63 pounds barley or bean meal, then cool and feed 5 to 28 pounds per day per animal depending on other feeds. From Patent Office, Report, 1844, pp. 387-88.

20. Prairie Farmer, June 1847, p. 181; Anderson, "Agriculture in Illinois," p. 94; Patent Office, Report, 1853, p. 6. "Stall" could possibly be a corruption of "stalk" feeding or derived from the sense of standing up to a feed bunk in an open lot. "Stall" feeding was also called the Kentucky or Ohio system.

21. McCoy, Historic Sketches, pp. 235-36, 238.

22. Prairie Farmer, Jan. 3, 1856, p. 2; ISAST, I:429-35.

23. ISAST, II:378, 376, 372-79.

24. Harris, Autobiography, pp. 74-79, 88-90; Prairie Farmer, June 1854, p. 239, July 1855, pp. 203, 207, Apr. 3, 1856.

25. Rammelkamp, Memoirs of John Henry, p. 70; Prairie Farmer, Nov. 1854, pp. 427-30; Bateman and Selby, Historical Encyclopedia, pp. 954-56.

26. McCoy, Historic Sketches, p. 236; Bateman and Selby, Historical Encyclopedia, pp. 806-7; [Ponting], Life, p. 28.

27. Prairie Farmer, Sept. 1846, p. 289, Feb. 1847, p. 57, Feb. 1849, pp. 48-49, June 1849, pp. 188-89, July 1852, p. 319, May 1853, p. 200, Dec. 11, 1856, Oct. 25, 1860, p. 258; ISAST, II:380-84; Patent Office, Report, 1851, pp. 439, 456; ibid., 1852, pp. 336, 340-41; ibid., 1854, p. 12; Anderson, "Agriculture in Illinois," p. 94; Pioneer, Northern Iowa, pp. 7-8; Porter, History of Harmon Township, J. Ill. State Hist. Soc. 10:609; Johnston, Sketches of the History of Stephenson County, Trans. Ill. State Hist. Soc., 1923, p. 317; Wistar, Autobiography, pp. 348-50; McCoy, Historic Sketches, pp. 235-36.

28. Prairie Farmer, Feb. 1844, p. 35, Jan. 1845, p. 18, Apr. 1848, p. 127, Mar. 1854, p. 92, Sept. 18, 1856, Jan. 28, 1858, Feb. 10, 1859, p. 82; Patent Office, Report, 1850, pp. 198-99; ibid., 1853, p. 336.

29. Prairie Farmer, Jan. 1845, p. 31, Mar. 1845, p. 65, Apr. 1845, p. 94, Aug. 1845, p. 187, Nov. 1846, p. 337, Jan. 1847, p. 29, Apr. 1853, p. 154; Rockford Register, Sept. 19, 1857, p. 2, Cole Notes.

30. Oliver, Eight Months in Illinois, pp. 103-5; Patent Office, Report, 1843, p. 121; Prairie Farmer, Feb. 1843, pp. 45-46, July 1843, p. 152, Nov. 1843, p. 261, Feb. 1844, p. 35, Dec. 1844, p. 287, Jan. 1845, p. 18, Oct. 1845, pp. 238-39, Sept. 1846, p. 289, Jan. 1847, p. 29, Feb. 1849, p. 64, Feb. 1851, p. 71, Jan. 1852, pp. 9, 45, Feb. 1852, pp. 66-67, June 1852, p. 263, Mar. 5, 1857, Oct. 15, 1857, Feb. 10, 1859, p. 83, May 19, 1859, p. 313, July 7, 1859, p. 8, Aug. 25, 1859, p. 115, Sept. 15, 1859, p. 161; Northwestern Farmer, May 5, 1858, pp. 166-67; Harris, Autobiography, pp. 88-90.

31. Buckingham, Early Settlers, J. Ill. State Hist.

Soc. 35:238; ISAST, III:148; <u>Beardstown Gazette</u>, Nov. 26, 1851, Cole Notes; Springfield <u>Daily Register</u>, Aug. 15, 1851, Cole Notes.

32. <u>Prairie Farmer</u>, Oct. 1847, p. 305, Jan. 1848, p. 27, June 1845, p. 159.

33. <u>Gem of the Prairie</u>, Nov. 15, 1850, quoted in Anderson, "Agriculture in Illinois"; Kerrick, Life and Character of Isaac Funk, pp. 161-65; <u>Prairie Farmer</u>, May 1850, p. 166.

34. Patent Office, Report, 1854, pp. 13-14; Newcombe, Alson J. Streeter, pp. 421-22; Johnston, "Sketches of the History of Stephenson County, pp. 288-90, 314-15, 317; Eli Strawn to Joel Hopkins, Sept. 20, Oct. 17, 25, 1860, Hopkins Papers.

35. <u>New York Tribune</u> in <u>Prairie Farmer</u>, Apr. 17, 1856; <u>Prairie Farmer</u>, Apr. 3, May 22, 1856, May 7 to June 25, 1857; Bateman and Selby, <u>Historical Encyclopedia</u>, p. 957.

36. <u>Prairie Farmer</u>, Nov. 6, 1856.

37. Cavanagh, <u>Funk of Funk's Grove</u>, pp. 52-53, 68, 78, 84; <u>Prairie Farmer</u>, Nov. 1848, pp. 336-37, Dec. 1849, p. 367, Dec. 1850, pp. 381-82, Jan. 1853, p. 20; <u>New York Tribune</u>, Apr. 25, 1854, p. 8.

38. <u>Cattleman</u> 34:110; <u>Prairie Farmer</u>, Oct. 1847, p. 305, Nov. 1848, p. 336, Dec. 1853, p. 467, Jan. 1855, p. 36, July 1855, p. 203, Aug. 1855, p. 263, May 1856, Feb. 17, 1859, p. 100; <u>Ottawa Free Trader</u>, Feb. 22, 1851, p. 2, Cole Notes; Democratic Press, <u>The Railroads</u>, p. 66.

39. <u>Prairie Farmer</u>, Feb. 1843, p. 44, Oct. 1847, p. 304, Nov. 1848, p. 336, Jan. 1852, p. 36, July 1855, p. 207, July 25, 1857; Patent Office, Report, 1850, p. 356; ibid., 1851, p. 448; ibid., 1852, pp. 336, 340-41, 343; ibid., 1853, p. 6; ibid., 1854, pp. 12-14; Anderson, "Agriculture in Illinois," p. 94; Bateman and Selby, <u>Historical Encyclopedia</u>, p. 806; <u>Quincy Whig</u>, Oct. 31, 1855, p. 3, Cole Notes.

40. ISAST, I:434-35; <u>Prairie Farmer</u>, July 1851, p. 317.

41. <u>Prairie Farmer</u>, Apr. 7, 1859, p. 216.

42. Ibid., Sept. 1851, p. 381, Aug. 1853, p. 322, July 9, 1857, Nov.-Dec. 1857; for the packing season, see the <u>Prairie Farmer</u> market columns, 1845-60.

43. Eighth Census, II, p. cxxxiii.

44. Patent Office, Report, 1861, p. 258.

Chapter 4

1. Pierce, <u>History of Chicago</u>, vol. 1, pp. 44-51, 127-31.

2. Seventeenth Census, I, p. 46.

3. Lee, Transportation, Ill. State Hist. Soc. 10:31-33; Pierce, History of Chicago, vol. 1, pp. 76-79.

4. Chicago Board of Trade, Report, 1880, p. 34; Clemen, American Livestock and Meat Industry, pp. 92-105; Prairie Farmer, June 1848, p. 199, Jan. 1853, p. 38.

5. Pierce, History of Chicago, vol. 1, pp. 137-41; Andreas, History of Chicago, vol. 1, p. 556.

6. Prairie Farmer, Sept. 1846, p. 296, Oct. 1846, p. 328; Illinois State Register, Nov. 12, 1849, p. 2, Cole Notes.

7. Prairie Farmer, Nov. 1848, pp. 336-37. Packed beef was cured meat salted in barrels, the normal way of preserving beef until the 1870s when packers put beef in sealed tin cans, hence tinned beef, and shipped refrigerated fresh quartered beef as dressed beef.

8. Ibid., Dec. 1849, p. 367; Cavanagh, Funk of Funk's Grove, pp. 49, 62, 66-67.

9. Prairie Farmer, Dec. 1849, p. 367.

10. Ibid., Dec. 1850, pp. 381-82; Clemen, American Livestock and Meat Industry, pp. 105-9.

11. Pierce, History of Chicago, vol. 1, p. 118; Lee, Transportation, pp. 33-35, 37.

12. Yungmeyer, Excursion into Early History, J. Ill. State Hist. Soc. 38:10-19; Throne, Burlington and Missouri Railroad, Palimpsest 33:1-25; Pierce, History of Chicago, vol. 2, pp. 41-43, 50; Agnew, Rock Island Railroad, Iowa J. Hist. 52:203-22; Lee, Transportation, p. 38; Hopkins, Economic History, p. 33; Gates, Illinois Central Railroad, pp. 86-98.

13. Chicago Board of Trade, Report, 1901, pp. 41, 169.

14. Prairie Farmer, Jan. 31, 1856, p. 18.

15. Patent Office, Report, 1854, p. 12; New York Tribune, Apr. 28, 1854, Ann. Iowa 27:296; Prairie Farmer, Apr. 3, May 22, 1856.

16. Prairie Farmer, July 1855, p. 207; Harper, Railroad and the Prairie, Ill. State Hist. Soc. Trans. 1923, pp. 106-8.

17. Prairie Farmer, Aug. 15, 1863, p. 99, Feb. 20, 1869, p. 64.

18. Ibid., Feb. 1851, p. 56, Jan. 1853, p. 20; U.S. Commissioner of Agriculture, Report, 1863, pp. 617-18, 1865, p. 86, hereafter this series will be cited as USDA Report; Chicago Board of Trade, Report, 1858, p. 26, 1860, p. 35, 1880, p. 35. There is disagreement among statistical sources

as to cattle receipts, shipments, and numbers packed or
butchered for city use before 1866. Figures differ because
some sources report calendar years and others report packing
season. Some sources even disagree internally; therefore, all
figures before 1866 are only approximate guides to the magnitude of the cattle business. In most cases, Chicago Board
of Trade figures are preferred here.

19. Chicago Board of Trade, Report, 1900, p. 45; USDA,
Report, 1862, p. 326.

20. Pierce, History of Chicago, vol. 2, pp. 98-99;
Griffiths' Review, p. 94. Plankington and Armour of Chicago
and Milwaukee opened a slaughter plant in Kansas City in
1869.

21. Griffiths' Review, pp. 19-22; Drovers' Journal,
June 28, 1939; Clemen, American Livestock and Meat Industry,
pp. 83-85.

22. Drovers' Journal, June 28, 1939; Pierce, History
of Chicago, vol. 2, pp. 92-94; ISAST, VI:314-24; Prairie
Farmer, Jan. 6, 1866, pp. 1-2.

23. U.S. Federal Trade Commission, Report on the Meat-
Packing Industry, pt. 3, pp. 194-95, hereafter cited as FTC,
Report.

24. Prairie Farmer, Apr. 3, 1873, p. 109, Mar. 10, Mar.
17, June 2, 1877; Union Stock Yards and Transit Company, Annual Report, 1889, pp. 5-7, hereafter this series will be
cited as Stock Yards, Report; Western Rural, Sept. 27, Nov.
22, 1879, Nov. 15, 1884, Dec. 20, 1890; Prairie Farmer, May
21, 1887, p. 329.

25. Stock Yards, Report, 1887, 1900.

26. Ibid., 1900, p. 31.

27. Pierce, History of Chicago, vol. 3, p. 204; Stock
Yards, Report, 1900, pp. 67-69.

Chapter 5

1. Chandler, Beginning of "Big Business" in American
Industry, Bus. Hist. Rev. 33:1-8.

2. Anderson, Refrigeration in America, pp. 24-26, 32-
35, 45-61, 91-92, 123, 142.

3. Leech and Carroll, Armour and His Times, pp. 125-
29; USDA Bureau of Animal Industry, Annual Report, 1884, pp.
265-66, hereafter this will be cited as BAI; Clemen, American Livestock and Meat Industry, p. 232; Prairie Farmer,
Mar. 11, 1876, p. 83.

4. Swift and Van Vlissingen, Yankee of the Yards, pp. 6, 126-30, 178, 185-90; FTC, Report on Private Car Lines, pp. 27-31; Kujovich, Refrigerator Car and the Growth of the American Dressed Beef Industry, Bus. Hist. Rev. 44:460-82; Armour, in Packers, Private Car Lines, and People, pp. 19-26, claims his father, Philip Armour, saw the need for refrigerator cars first rather than Swift, but other sources do not support that assertion.

5. Swift and Van Vlissingen, Yankee of the Yards, pp. 27-30, 67-69, 130-32, 137, 205; FTC, Report on Meat Packing, pt. 1, p. 214.

6. Prairie Farmer, Nov. 4, 1882, p. 4; USDA, Report, 1876, p. 320; Nimmo, Report on the Internal Commerce of the United States, H. Exec. Doc. 7, pt. 3, pp. 157-58; Chicago Board of Trade, Report, 1886-1900; Pierce, History of Chicago, vol. 3, pp. 110-11; BAI, Report, 1884, p. 267; Prairie Farmer, Oct. 6, 1883, p. 640.

7. Clemen, American Livestock and Meat Industry, pp. 228-31, 237.

8. Prairie Farmer, Dec. 9, 1882, p. 9.

9. Ibid., Apr. 14, 1883, p. 228; Aldrich, Wholesale Prices, Wages, and Transportation, S. Rept. 1394, pt. 1, p. 525.

10. Prairie Farmer, Aug. 14, 1883, p. 228; Aldrich, Wholesale Prices, Wages, and Transportation, pt. 1, pp. 525-26; Kujovich, Refrigerator Car and the Growth of the American Dressed Beef Industry, pp. 460-82.

11. Drovers' Journal, Oct. 7, 1882, p. 4; Swift, Yankee of the Yards, pp. 67-69; Prairie Farmer, Dec. 9, 1882, p. 9; Western Rural, Sept. 18, 1886, p. 604; Butchers' Advocate, Dec. 21, 1890, p. 11, Jan. 28, 1891, p. 3; Clemen, American Livestock and Meat Industry, pp. 243-44; U.S. Congress, H. Exec. Doc. 267, p. 176; Kujovich, Refrigerator Car and the Growth of the American Dressed Beef Industry, pp. 466-68.

12. Swift v. Sutphin; In Re Barber; Minnesota v. Barber; Clemen, American Livestock and Meat Industry, pp. 245-51; Butchers' Advocate, Nov. 21, 1894, p. 9.

13. Leech and Carroll, Armour and His Times, pp. 51-52; Tenth Census, III, p. 1109; Pierce, History of Chicago, vol. 3, pp. 115-18; BAI, Report, 1884, pp. 262-64; U.S. Congress, H. Exec. Doc. 267, pp. 174-76.

14. Nimmo, Report on the Internal Commerce of the United States, pp. 286-88.

15. Zimmerman, Live Cattle Export Trade, Agr. Hist.

36:46-47; Griffiths' Review, 1876, p. 5.

16. Morris, Heritage from My Father, p. 209; Nimmo, Report on the Internal Commerce of the United States, pp. 266-67; Zimmerman, Live Cattle Export Trade, p. 47; North British Agriculturalist quoted in Prairie Farmer, Nov. 2, 1878, p. 349.

17. Prairie Farmer, Feb. 9, 1878, p. 45, June 15, 1878, p. 189, Aug. 3, 1878, p. 246, Oct. 26, 1878, p. 341, Nov. 2, 1878, p. 349, Apr. 26, 1879, p. 133.

18. Ibid., Oct. 16, 1880, p. 333; Chicago Board of Trade, Report, 1900, p. xxxvii; contrary to the impression in Zimmerman, exports of live beef cattle to England did not decline greatly after 1885. See Perrin, North American Beef and Cattle Trade, Econ. Hist. Rev. 24:431-32.

19. Western Farm Journal, Apr. 5, 1879, pp. 1-2; Iowa Homestead, Jan. 10, 1890, p. 3, Apr. 4, 1890, p. 3, Apr. 11, 1890, p. 5, Apr. 18, 1890, p. 1, May 8, 1891, p. 436; Prairie Farmer, Nov. 25, 1876, p. 380, citing Dundee Advertiser.

20. Prairie Farmer, Nov. 2, 1878, p. 349, Feb. 9, 1878, p. 48, July 27, 1878, p. 237; Griffiths' Review, 1877, pp. 5, 120.

21. Prairie Farmer, Oct. 15, 1880, p. 333; Chicago Board of Trade, Report, 1900, pp. xxxvii, 161; USDA, Yearbook, 1894, p. 10, 1906, p. 256; for an early British view of the dressed beef trade, see MacDonald, Food from the Far West, pp. 253-61, 277, 284-89; Perrin, North American Beef and Cattle Trade, pp. 434-44; Chicago Tribune, June 17, 1964, sec. 3, p. 7 under the heading "Beef from U.S. Bids for British Market," carried an article about an attempt of the American Meat Institute to sell U.S. packaged beef in England by opening a display of 7,000 pounds of beef from Cedar Rapids, Iowa, at the Smithfield market in London. The promoters hoped to develop a "selective trade" in American beef over the long term if favorable Atlantic freight rates could be arranged and if the British bought.

22. FTC, Report, pt. 2, pp. 13-17.

23. U.S. Congress, Report of the Select Committee on the Transportation and Sale of Meat Products, S. Rept. 829, pt. 1, pp. 1-20; Iowa Homestead, Oct. 30, 1891, p. 1007.

24. East, "Distillers' and Cattle Feeders' Trust," J. Ill. State Hist. Soc. 45:104-5; Peoria Board of Trade, Annual Report, 1880-1884; Gressley, Bankers and Cattlemen, p. 66; "Nelson Morris," Dictionary of American Biography, 13:217-18.

25. FTC, Report, Summary, pp. 33-35.

26. Ibid., pt. 2, pp. 43; U.S. Commissioner of Corporations, Report on the Beef Industry, H. Doc. 382, p. 7; IDAT, 1898, p. 300.

Chapter 6

1. Dale, Range Cattle Industry, tells the story of the growth, overexpansion, collapse, and reorganization of the range industry; Gressley, Bankers and Cattlemen, concerns the 1880s and 1890s; see also Prairie Farmer, July 14, 1883, p. 436, warning of overexpansion and speculation in range cattle.

2. Iowa City Republican, quoted in Ross, Iowa Agriculture, p. 72; Ross, Iowa Agriculture, pp. 71-72; Anderson, "Agriculture in Illinois," p. 267; Prairie Farmer, Nov. 28, 1861, p. 353, Jan. 4, 1863, p. 1, Apr. 16, 1864, p. 268, May 6, 1865, p. 356, June 24, 1865, p. 505, Oct. 7, 1865, p. 268, Dec. 16, 1865, p. 431, Apr. 27, 1872, p. 133.

3. Prairie Farmer, Dec. 12, 1863, p. 386, May 6, 1865, p. 356, Dec. 15, 1865, p. 431.

4. Prime, Model Farms, pp. 430-34.

5. MacDonald, Food from the Far West, pp. 143-44; Gates, Cattle Kings, Miss. Valley Hist. Rev. 35:395.

6. Prairie Farmer, Feb. 27, 1870, p. 64, Feb. 17, 1872, p. 56, Oct. 23, 1875, p. 344; MacDonald, Food from the Far West, pp. 120-22; Prime, Model Farms, pp. 282-84, 501, 503-4.

7. Prairie Farmer, Dec. 6, 1862, p. 353; MacDonald, Food from the Far West, pp. 140-42.

8. Bogue, Prairie to Corn Belt, p. 99; MacDonald, Food from the Far West, pp. 120-22; Prairie Farmer, Feb. 3, 1877, p. 33, Dec. 22, 1877, p. 408, Nov. 2, 1878, p. 349, Jan. 16, 1886, p. 366, Dec. 30, 1882, pp. 1-2; Johnson, E. S. Ellsworth, Ann. Iowa 35:24; Hopkins, Economic History, p. 224 n175.

9. Hopkins, Economic History, p. 181.

10. Iowa Homestead, Sept. 18, 1891, p. 862; Prairie Farmer, Feb. 15, 1878, p. 53.

11. Anderson, "Agriculture in Illinois," pp. 273-75; Baldwin, Driving Cattle from Texas to Iowa, Ann. Iowa 14:243-62.

12. ISAST, VII:411, 426; Anderson, "Agriculture in Illinois," pp. 275-76.

13. ISAST, VII:135-36; Prairie Farmer, Aug. 7, 1869, p. 249.

14. Illinois State Journal, Feb. 23, 1870, p. 3, Cole Notes; Prairie Farmer, Nov. 26, 1870, p. 369.

15. See McCoy, Historic Sketches, for the story of Abilene, McCoy's comments about Texas fever, and his battle against the state restrictions on Texas cattle. Dykstra, Cattle Towns, pp. 17-30, is the proper antidote to McCoy's view of his actions.

16. Prairie Farmer, Feb. 23, 1867, p. 114, Apr. 13, 1867, p. 234; USDA, Report, 1868, pp. 38-40; Prairie Farmer, Aug. 1, 1868, p. 36, Aug. 8, 1868, p. 44, Aug. 15, 1868, p. 50.

17. McCoy, Historic Sketches, pp. 221-29.

18. ISAST, VII:134-57; Iowa State Agricultural Society Report, 1872, passim, hereafter this series will be cited as Iowa SASR.

19. Report of the convention in ISAST, VII:173-273.

20. Bierer, Short History of Veterinary Medicine, p. 54.

21. Ibid., pp. 49-55, 60; USDA, Yearbook, 1956, pp. 2-4, 269; by 1940 the cattle tick had been eliminated from all of the United States through a federally enforced dipping program, except for a narrow strip of land in Texas bordering Mexico.

22. McCoy, Historic Sketches, pp. 222, 225.

23. Prairie Farmer, Mar. 2, 1867, p. 129; Illinois Revised Statutes, 1874, pp. 141-44; McCoy, Historic Sketches, p. 254.

24. Prairie Farmer, Aug. 17, 1872, p. 270, Aug. 24, 1872, p. 268, Jan. 11, 1873, p. 12, Sept. 30, 1876, p. 316, Sept. 26, 1885, p. 626; Illinois Board of Live Stock Commissioners, Annual Report, 1890-1900, passim.

25. Prairie Farmer, Aug. 15, 1879, p. 261; ISAST, VIII:252-53; Breeders' Gazette, May 13, 1886, p. 697; Prairie Farmer, July 27, 1872, p. 236, Feb. 25, 1888, Dec. 31, 1887, p. 851; Gates, Cattle Kings, p. 398.

26. Prairie Farmer, May 15, 1886, p. 310.

27. Ibid., Jan. 6, 1877, p. 8.

28. Ibid., July 25, 1868, p. 26.

29. USDA, Report, 1870, pp. 350-51; IDAT, XIV:105.

30. Prairie Farmer, Feb. 10, 1872, p. 48.

31. USDA, Report, 1876, p. 319.

32. Prairie Farmer, Dec. 1, 1877, p. 381; IDAT, XVII: 463, XVIII:470-71.

33. _Prairie Farmer_, Oct. 4, 1884, p. 628; Tenth Census, III, pp. 978, 991, 1000, 1005, 1008, 1011; Drovers' Journal, _Year Book_, 1922, p. 36; Dale, _Range Cattle Industry_, passim.

34. Address by Governor John Hamilton of Illinois at the Fat Stock Show, Nov. 1883, in IDAT, XXI:106-7.

35. _Iowa Register_ in _Prairie Farmer_, Oct. 4, 1884, p. 628; _Iowa Homestead_, Mar. 27, 1891, p. 289, Sept. 18, 1891, p. 860.

36. Cavanagh, _Funk of Funk's Grove_, pp. 49, 66-67; see Chapter 3, pp. 23-24.

37. Loan ledgers in the Hopkins Papers show that a considerable amount of money was being loaned to neighboring farmers in the 1880s and 1890s.

38. McCoy, _Historic Sketches_, p. 242; Bogue, Pioneer Farmers and Innovation, _Iowa J. Hist._ 56:26.

39. _Prairie Farmer_, Sept. 19, 1896, p. 3.

40. Hopkins, _Economic History_, pp. 33, 40, 143-52, 225, n182, n183; _Prairie Farmer_, Feb. 1, 1890, p. 65; Stock Yards, Report, 1898, p. 10, 1900, pp. 68-71; USDA, _Yearbook_, 1918, pp. 101-8.

41. U.S., Statutes at Large, 17:584-85; Hopkins, _Economic History_, p. 171.

42. Pierce, _History of Chicago_, vol. 3, pp. 137-39; Hopkins, _Economic History_, pp. 171, 230-32; _Prairie Farmer_, Oct. 26, 1872, p. 337; Aldrich, Wholesale Prices, Wages, and Transportation, pt. 1, pp. 525-27; U.S. Congress, H. Doc. 855, vol. 144, pp. 128-33; ibid., S. Rept. 829, pp. 1-4, 18-20; Grodinsky, _Iowa Pool_, passim.

43. _Prairie Farmer_, market columns, 1880s, 1890s; Clemen, _American Livestock and Meat Industry_, pp. 635-39.

44. _Prairie Farmer_, Jan. 27, 1900; _Iowa Homestead_, Dec. 18, 1891, p. 1173.

45. Illinois State Board of Agriculture, Statistical Report, 1883-1900, passim; Murray, Shipping the Fat Cattle, _Palimpsest_ 28:85-93.

46. _Prairie Farmer_, Oct. 27, 1866, p. 273, Nov. 17, 1866, p. 321.

47. _Western Rural_, Feb. 23, 1884, p. 124, Mar. 1, 1884, p. 136, Nov. 15, 1884, p. 732; _Prairie Farmer_, Nov. 3, 1866, p. 289; Sales receipts of R. Waugh, Commission Dealers, 1897-1900, Hopkins Papers.

48. Clemen, _American Livestock and Meat Industry_, pp. 485-89.

49. _Prairie Farmer_, Nov. 18, 1865, p. 373, Mar, 9, 1867, p. 157, Apr. 23, 1870, p. 128; MacDonald, _Food from_

the Far West, pp. 273, 277, 279; Griffiths' Review, 1876, pp. 21-23.

50. Prairie Farmer, Jan. 17, 1874, p. 24.

51. Ibid., Sept. 20, 1884, p. 605.

52. USDA, Report, 1883, pp. 281-82; Prairie Farmer, July 31, 1886, p. 495; BAI, Report, 1887-88, pp. 363, 410-13; USDA, Yearbook, 1894, p. 12.

53. Iowa Homestead, Jan. 9, 1891, p. 25; Prairie Farmer, Jan. 25, 1896, p. 2, Feb. 15, 1896, p. 1, Apr. 4, 1896, p. 7, July 4, 1896, p. 6.

54. Ledgers and commission house receipts, 1897-1900, Hopkins Papers.

55. Field and Farm, Oct. 6, p. 4, Oct. 13, p. 9, Oct. 27, p. 4, Nov. 24, p. 9, Dec. 15, 1888, p. 9; Butchers' Advocate, Apr. 8, 1891, p. 3; Drovers' Journal, Year Book, p. 36; Hopkins, Statistical Study, Iowa State Coll. Agr. Exp. Sta. Bull. 101, pp. 344-53, 359-61.

56. Prairie Farmer, Aug. 10, p. 256, Sept. 14, 1872, p. 296, Jan. 11, p. 13, Feb. 8, 1873, p. 48; Field and Farm, Jan. 7, p. 4, Jan. 28, 1888, p. 4; Wallaces' Farmer, Feb. 3, p. 85, Feb. 17, 1899, p. 125.

57. Prairie Farmer, Aug. 29, 1896, p. 6; Drovers' Journal, Year Book, 1922, p. 36; ibid., 1936, p. 10.

58. Iowa Homestead, Mar. 20, 1891, p. 265; Prairie Farmer, May 22, 1886, p. 326; Illinois Farmer's Institute, Annual Report, pp. 261-64; University of Illinois, Stock Feeding, Bull. 36, pp. 421, 426; BAI, Report, 1885, p. 376.

59. Prairie Farmer, May 18, 1888, p. 319; Iowa Homestead, Nov. 6, 1891, p. 1030; Illinois Farmer's Institute, Report, 1899, pp. 331-33, 337.

60. Iowa Homestead, Nov. 6, 1891, p. 1037.

61. U.S. Congress, S. Report 829, pp. 1-4, 6-7, 10-11.

62. Stock Yards, Report, 1903, pp. 7-9; Prairie Farmer, 1890-1900, passim.

63. Illinois Farmer's Institute, Report, 1897, pp. 223-24.

Chapter 7

1. This discussion is based on statistics taken from the U.S. Census 1850-1900 for the number of cattle on farms, and from Hopkins, Economic History, pp. 71-80. In 1900, "other" cattle was taken to be all but dairy cows two years

old or over. The author is aware of faults in this system and of other methods of indicating relative concentrations of cattle, as well as some problems in the accuracy of the printed census. Although attempts have been made to have more accurate figures for cattle in the census years for the nation and the states, any attempt to look at counties is dependent on the "uncorrected" census.

2. Prairie Farmer, July, 1855, p. 207.

3. Hawk in "Iowa Farming Types" delineated Iowa types of farming by "other" cattle per farm and came to some different conclusions. His 1880 map included a high-density area of beef cattle in the northwest which centered on Palo Alto and Pocahontas counties, which does not appear on my maps. He had an area of concentration in eastern Iowa that corresponded with my beef cattle concentrations and Bogue's corn and swine concentrations. Hawk's study ended with 1880. Bogue kindly supplied me with maps of some of his data on leading counties for corn, swine, and milk cows in Illinois and Iowa from 1850 to 1890. Bogue measured corn per improved acre and swine and milk cows per farm on a county basis so that comparisons are not exact. He found a closer relationship between leading corn and swine producing counties than there appears to be between my leading cattle counties and his leading corn and swine counties. For 1880 and 1890 there was a closer relationship between leading cattle and hog counties than between leading cattle and corn counties. There was a close correspondence between heavy numbers of milk cows and beef cattle densities in 1860 in northeastern urban Illinois, and there was some correspondence between the two in 1860 in northwestern Illinois, but not in northeastern Illinois (heavy number of milk cows, light number of beef cattle). Otherwise, in regard to milk cow-beef cattle densities, there were no remarkable correspondences.

There were a number of ways to map these statistics. Bogue argued that beef cattle per farm unit was better than per 100 acres. The former would indicate immediate intensity of endeavor perhaps, but is open to certain distortions too. One could also map beef cattle per improved acre, but that would produce maps showing densities of cattle in grazing counties where total cattle numbers might still be less than in counties where corn was the major feed. The decision to use beef cattle per hundred acres was arbitrary and its limits are recognized. It is offered as an approximate indication of cattle feeding areas. The delineation of these areas, in general, seems to be supported by information from other sources, though the delineation does not necessarily indicate the real relative magnitudes of cattle feeding in the counties. All these calculations are dependent upon census figures which are undergoing "correction" by scholars, although so far only for state totals and not the county totals necessary here.

4. *Prairie Farmer*, Apr. 19, 1847, May 29, 1869, p. 169, June 12, 1875, p. 192, Feb. 23, 1878, p. 61, May 11, 1878, p. 152, Apr. 7, 1883, p. 217, May 17, 1884, p. 307; ISAST, VIII:209; IDAT, XI:109, XIV:125, 174; Anderson, "Agriculture in Illinois," pp. 290-91; Prime, *Model Farms*, p. 277.

5. University of Illinois, Stock Feeding, Bull. 36, p. 1; Newlove, "Economic History of Illinois Agriculture," pp. 81-82; Bogue, Swamp Land Act, *Agr. Hist.* 25:169-80; *Prairie Farmer*, Jan. 15, 1869, p. 17, June 4, 1870, p. 172; Smith, "Historical Geography of Champaign County, Illinois," pp. 112-13, 129-30.

6. This would seem true for east-central Illinois where, by 1900, most of the ten counties with the highest percentage of tenants (54% to 62%) were located. Only one of those counties ranked as high as fifth in the nine levels of cattle density on the 1900 map. The evidence for Iowa is less conclusive because of the ten counties with the highest percentage of tenants (45% to 52%); three each were ranked at the sixth, seventh, and eighth levels of cattle density. One county ranked at the fifth level. However, even the county with the highest percentage of tenancy in Iowa ranked below the eleventh highest Illinois county so that the level of tenancy in Iowa was lower than in Illinois in 1900. Twelfth Census, V, pp. 1, 73-75, 79-81.

7. Gates, Cattle Kings, *Miss. Valley Hist. Rev.* 35: 399-406; idem, Frontier Landlords and Pioneer Tenants, *J. Ill. State Hist. Soc.* 38:178-89, 204; Scully Papers.

8. Allerton, *Systematic Farming*, passim.

9. Sanders, James N. Brown, *Trans. Ill. State Hist. Soc.*, 1912, pp. 172-75; *Prairie Farmer*, Sept. 3, 1864, p. 146, June 30, 1877, p. 205, Sept. 4, 1880, p. 285; ISAST, I:429-36, II:372-78.

10. McClelland, Jacob Strawn, *J. Ill. State Hist. Soc.* 34:201-5.

11. Ibid., pp. 182-95.

12. Rammelkamp, Memoirs of John Henry, *J. Ill. State Hist. Soc.* 18:70.

13. Bateman and Selby, Historical Encyclopedia, p. 976.
14. Ibid., p. 777.
15. Ibid., pp. 793-94.
16. History of Morgan County, pp. 537-66.
17. Little, "Early Days of Henry County," in Little Papers; Bogue, Prairie to Corn Belt, pp. 193-215, 237-38; Hawk, "Iowa Farming Types," pp. 92-98.
18. Wallaces' Farmer, Sept. 1, 1899, p. 701; Prairie Farmer, Feb. 28, 1880, p. 69; Anderson, "Agriculture in Illinois," pp. 287-89; Prime, Model Farms, p. 262; Prairie Farmer, Dec. 6, 1873, p. 392, Nov. 15, 1872, p. 368; IDAT, XIII:193.
19. Prairie Farmer, Aug. 10, 1872, p. 256, Sept. 14, 1872, p. 296, Jan. 11, 1873, p. 13; Wallaces' Farmer, Sept. 1, 1899, p. 701.
20. Brinkman, "Historical Geography of Improved Cattle," Appendix, maps 14, 24, 25.
21. See Maps 7-12; Bogue, Prairie to Corn Belt, pp. 12, 220-23.
22. Iowa SASR, 1875, p. 438; Hopkins, Economic History, p. 107; Prairie Farmer, Feb. 3, 1877, p. 33; Bogue, Prairie to Corn Belt, pp. 229-30.
23. Bushnell, Iowa Resources and Industries, p. 59.
24. Prairie Farmer, May 1848, p. 164, Apr. 12, 1878, p. 120; USDA, Report, 1873, p. 392.
25. Prairie Farmer, Mar. 28, 1874, p. 104.
26. Iowa Homestead, Jan. 30, 1891, p. 101.
27. Bogue, Prairie to Corn Belt, p. 96.
28. Ibid., pp. 97-98.
29. Bogue, Pioneer Farmers, Iowa J. Hist. 55:24-25.
30. Prime, Model Farms, pp. 430-34.
31. Prairie Farmer, Nov. 1854, pp. 427-30.
32. Harris, Autobiography, Trans. Ill. State Hist. Soc., 1923, pp. 90-91; Illinois State Register, Mar. 24, 1853, p. 1; Urbana Union, Nov. 10, 1853, p. 2, Cole Notes.
33. USDA, Report, 1866, p. 319.
34. McCoy, Historic Sketches, p. 243.
35. Prairie Farmer, Dec. 23, 1876, p. 413.
36. National Live Stock Association, Proceedings, p. 177.
37. Conway, Cattle Handbook; Wallaces' Farmer, Mar. 3, 1899, p. 175.
38. Prime, Model Farms, p. 47.
39. Wentworth, Observations, Agr. Hist. 27:41, 43.
40. Ibid., p. 42.

Chapter 8

1. Bardolph, _Agricultural Literature_, passim; Ayer & Son's, _American Newspaper Annual_, 1880, 1891, 1900; Gates, _The Farmer's Age_, pp. 338-58; Ogilvie, _Pioneer Agricultural Journalists_, pp. 36-37.

2. Ross, _Iowa Agriculture_, pp. 35-36, 88-91; Lord, _Wallaces of Iowa_, pp. 83, 92, 120, 130-31.

3. _Prairie Farmer_, Jan. 1845, p. 20, Sept. 1850, p. 265, Nov. 1850, p. 349.

4. Hamilton (Iowa) _Freeman_, July 15, 1858, quoted in Bogue, _Prairie to Corn Belt_, p. 196; _Prairie Farmer_, July 1847, p. 201.

5. _Prairie Farmer_, Dec. 5, 1861, p. 374, Mar. 10, 1877, p. 73, Sept. 1, 1877, p. 273, July 17, 1880, p. 229, Oct. 4, 1884, p. 628.

6. Ibid., Sept. 1851, p. 421, May 1852, pp. 243-44, July 1852, p. 339, May 1853, pp. 176-77, June 1853, pp. 223-34, July 1853, pp. 264, 279, Jan. 10, 1856; _Iowa Homestead_, Aug. 13, 1869, p. 2; _Drovers' Journal_, Sept. 12, 1882, p. 2.

7. _Prairie Farmer_, June 6, 1861, p. 368, Nov. 29, 1862, p. 345, Jan. 26, 1867, p. 56, Oct. 3, 1868, p. 105, July 31, 1869, p. 241, Oct. 7, 1871, p. 313, July 15, 1876, p. 229, Dec. 8, 1877, p. 389, Dec. 15, 1877, p. 397, Dec. 29, 1877, p. 413, Dec. 8, 1883, p. 780, Oct. 4, 1884, p. 628, and passim; _National Livestock Journal_, Mar. 1885, pp. 110-11.

8. _Iowa Homestead_, _Wallaces' Farmer_, _Prairie Farmer_, passim; _Western Agriculturalist_, June 1886, p. 4; _Prairie Farmer_, May 19, 1859, p. 307, Jan. 1845, p. 18.

9. _Prairie Farmer_, Jan. 1845, p. 18, June 1847, pp. 183, 215, Nov. 1846, p. 337, Dec. 1850, p. 368, Nov. 17, 1883, p. 729; _Northwestern Farmer_, Feb. 1858, p. 52.

10. _Northwestern Farmer_, Jan. 1856, p. 9.

11. _Prairie Farmer_, Aug. 1851, p. 355.

12. Ibid., Dec. 1853, p. 447.

13. Ibid., Oct. 1855, p. 303.

14. Ibid., Aug. 1851, p. 359.

15. Ibid., Apr. 18, 1874, p. 125.

16. For two examples, see ibid., Feb. 15, 1862, p. 100, Apr. 13, 1867, p. 235.

17. Ibid., Dec. 10, 1864.

18. Ibid., Nov. 17, 1866, p. 314, July 4, 1868, p. 2, Feb. 25, 1871, p. 64, Sept. 1, 1877, p. 273.

19. Ibid., Jan. 24, 1862, p. 51, Oct. 29, 1870, p. 340, Feb. 5, 1870, p. 33, Jan. 15, 1887, p. 36; Field and Farm, Jan. 15, 1887, p. 5; Western Agriculturalist, June 1886, p. 11.

20. See for example: Prairie Farmer, Aug. 3, 1872, p. 245, Oct. 17, 1874, p. 333, Jan. 2, 1875, p. 5, July 3, 1875, p. 213, Apr. 14, 1877, p. 117, Sept. 22, 1883, p. 596, Nov. 17, 1883, p. 721, Apr. 19, 1884, p. 244, Sept. 6, 1884, p. 562, Mar. 21, 1885, p. 180; Western Rural, Oct. 30, 1886, p. 707; Iowa Homestead, Jan. 1, 1890, p. 2, Jan. 31, 1890, p. 4, Feb. 2, 1890, p. 4, Feb. 28, 1890, p. 2; Wallaces' Farmer, Jan. 1, 1899, pp. 24, 25.

21. Prairie Farmer, Feb. 15, 1884, pp. 100-101, July 26, 1884, p. 468, Aug. 9, 1884, p. 500.

22. Field and Farm, Feb. 12, 1887, p. 4, and passim, 1887-1888.

23. Prairie Farmer, Jan. 1846, pp. 16-17.

24. Ibid., Jan. 1849, p. 17.

25. Ayer and Son's, American Newspaper Annual, 1880, p. 421, 1891, pp. 1182-83, 1900, p. 1356; Bardolph, Agricultural Literature, p. 170; circulation figures were guarded items among agricultural papers because they contested among themselves as to which had the most readers.

26. Prairie Farmer, Sept. 12, 1863, p. 161.

27. ISAST, I:21; Iowa SASR, 1874, pp. 485 ff. for the founding meetings.

28. ISAST, passim, 1854-1900; Iowa SASR, passim, 1854-1900.

29. Bardolph, Agricultural Literature, p. 71.

30. ISAST, II:vii-ix; Iowa SASR, 1857, pp. 448-50; Beinhauer, County, District, and State Agricultural Societies, Ann. Iowa 20:50-55; Prairie Farmer, Feb. 2, 1860, p. 68; Ross, Iowa Agriculture, pp. 84-86; IDAT, XXXVIII:297.

31. Prairie Farmer, Nov. 1850, pp. 330, 334.

32. Ibid., Jan. 28, 1858, p. 85; Northwestern Farmer, Apr. 1858, pp. 130-31; Iowa Farmers' Advocate, Dec. 1847, p. 85 quoted in Throne, "Book Farming" in Iowa, Iowa J. Hist. 49:118; Hildreth, Life and Times, pp. 180-83; practically any issue of the Prairie Farmer in September, October, November, and December up to 1875 will have some reports on fairs in Illinois and Iowa. The state agricultural society reports also carried county summaries, which frequently told of the fairs. After the middle 1870s, the Illinois Transactions contained only a statement of receipts and expenses for

the county societies, but the Iowa Reports continued to have information about local fairs and conditions.

33. *Iowa Homestead*, Feb. 26, 1869, p. 60.

34. *Prairie Farmer*, Dec. 5, 1874, p. 389, Dec. 12, 1874, p. 394, Feb. 24, 1877, p. 60.

35. Ibid., Sept. 19, 1868, p. 89.

36. IDAT, 1878-1900, passim.

37. Ibid., XVI:111.

38. Ibid., pp. 67, 79-82.

39. *Prairie Farmer*, Dec. 14, 1878, p. 397.

40. IDAT, XVII:95, 142-45, 147.

41. *Prairie Farmer*, Mar. 27, 1880, p. 101; IDAT, XVIII: 77.

42. IDAT, XVIII:71, XIX:93.

43. Ibid., XIX:5.

44. *Prairie Farmer*, Mar. 24, 1883, p. 180; Illinois Department of Agriculture, *Report on the 8th Fat Stock Show*, pp. 250-53.

45. *Prairie Farmer*, Mar. 24, 1883, p. 180.

46. Ibid., Dec. 8, 1883, p. 785.

47. Ibid., Nov. 22, 1884, p. 737.

48. IDAT, XIX:205-6.

49. Ibid., XXIII:8-9.

50. Ibid., pp. 27-28.

51. *Prairie Farmer*, Nov. 20, 1886, p. 761.

52. *Field and Farm*, Jan. 22, 1887, p. 1.

53. IDAT, XXIV:152, XXV:6.

54. Ibid., XXV:147, 216, XXVI:6; *Prairie Farmer*, Dec. 29, 1888, p. 846.

55. IDAT, XXXV:33-35, 76-77.

56. Union Stock Yards and Transit Company, Review, pp. 10-16.

57. IDAT, XXIII:132, 135-39, 146-51.

58. Ibid., pp. 290, 304.

59. Ibid., XXVII:160. See Van Stuyvenberg, *Margarine*, or Hayter, *Troubled Farmer*, ch. 4.

60. Ritzman, Baby Beef, USDA Circ. 105, pp. 3, 31-32.

61. ISAST, II:317-18.

62. *Prairie Farmer*, Oct. 1853, p. 394; see also ibid., Oct. 1843, p. 217, Nov. 1843, p. 245, Dec. 1851, pp. 561-64, Feb. 1852, p. 96, Aug. 1853, p. 229, Nov. 1853, pp. 424-25, Mar. 1854, pp. 89-90, Aug. 1854, p. 288, Jan. 1855, p. 32, Nov. 1855, p. 350, Oct. 17, 1863, p. 241; ISAST, I:84, 108, 133, II:xiii.

63. USDA, Report, 1873, p. 392.

64. Prairie Farmer, Oct. 4, 1873, p. 317.

65. Brinkman, "Historical Geography of Improved Cattle," pp. 106-12, 126-28, 183-204, appendix, pp. 72, 74-78; Prairie Farmer, Jan. 15, 1857.

66. Brinkman, "Historical Geography of Improved Cattle," appendix, maps 1, 8.

67. ISAST, V:372.

68. USDA, Report, 1873, p. 392.

69. ISAST, III:301-4.

70. Bogue, Prairie to Corn Belt, p. 88; Ross, Iowa Agriculture, pp. 45-46; Northwestern Farmer, Nov. 1857, p. 420.

71. ISAST, II:xiii, 64-67, III:177, IV:123-28, 254-59, 274, 277, 280, 285, 292, 300, V:111-14, 263-65, 290-92, VI: 26-28, 50, 114-16.

72. Brinkman, "Historical Geography of Improved Cattle," appendix, pp. 91-92, 95, 97-98, 100-105, 107-8, 112-14, maps 14, 15, 26.

73. Prime, Model Farms, pp. 154, 280.

74. Iowa Homestead, Aug. 13, 1869, p. 2.

75. IDAT, XX:462-67; Prairie Farmer, Dec. 2, 1882, p. 4, Feb. 24, 1883, p. 116, May 26, 1883, p. 324.

76. Prairie Farmer, May 26, 1883, p. 324.

77. Bogue, Prairie to Corn Belt, pp. 90-91.

78. Beinhauer, Early Iowa Agricultural Organizations, Ann. Iowa 34:422-24; Prairie Farmer, Jan. 12, 1884, p. 20; Iowa Homestead, Dec. 11, 1891, p. 1153.

79. Illinois Farmer's Institute, Report, 1896, p. 176; IDAT, XXXVI:282-83.

80. Bogue, Prairie to Corn Belt, pp. 90-91; Eleventh Census, VI, pp. 283-85.

81. Prairie Farmer, Apr. 23, 1870, p. 128; Iowa Homestead, Jan. 14, 1870, in Hopkins, Economic History, p. 212 n62.

82. IDAT, IX:203-4, XV:202, XIX:225, XX:426-29; Prairie Farmer, Mar. 28, 1874, p. 104, Feb. 19, 1876, p. 64, Apr. 8, 1876, p. 120, May 12, 1877, p. 152, Nov. 10, 1883, p. 713; Prime, Model Farms, pp. 45-47, 154, 429-30, 444-45, 453-54; IDAT, XVII to XXIII, printed a list of all the county fairs and tabulated for each fair the number and kind of purebred stock exhibited.

83. Prairie Farmer, Feb. 19, 1876, p. 64.

84. Ibid., Jan. 8, 1876, p. 16.

85. Ibid., May 21, 1881, p. 165.

86. Iowa Homestead, July 9, 1869, p. 210, Apr. 4, 1890, p. 5.

87. Prairie Farmer, May 6, 1876, p. 149, Aug. 27, 1881, p. 277.

88. Prime, Model Farms, pp. 444-45; and for general comments urging improvement see ibid., pp. 154, 280; Prairie Farmer, Apr. 27, 1872, p. 133, May 6, 1876, p. 149, and passim to 1900; IDAT, XVII:202, XX:426-29; Bushnell, Iowa Resources, pp. 59-61.

89. USDA, Report, 1868, pp. 515-17, 1869, p. 542; Hopkins, Economic History, p. 72; Prairie Farmer, May 3, 1879, p. 141.

90 USDA, Report, 1868, pp. 515-16; Prairie Farmer, Dec. 11, 1875, p. 397; Wallaces' Farmer, Sept. 29, 1899, p. 786.

91. Prairie Farmer, May 3, 1879, p. 141, and passim 1880-1890.

92. IDAT, XVI:98.

93. MacDonald, Food from the Far West, pp. 146-48.

94. IDAT, XVII:134.

95. Illinois State Board of Agriculture, Report of the American Fat Stock and Dairy Show, pp. 236-39, 248-53. This report had comparative tables for all eight shows from 1878 to 1885.

96. Ibid., p. 263.

97. Prairie Farmer, Nov. 20, 1880, p. 393, Dec. 4, 1880, p. 389; National Livestock Journal, Mar. 1885, pp. 110-11, Apr. 1885, pp. 155-56, Dec. 1885, p. 500.

98. Breeders' Gazette, Mar. 18, 1886, p. 375.

99. Iowa Homestead, Jan. 9, 1891, p. 25, July 31, 1891, p. 706; Prairie Farmer, Apr. 19, 1890, p. 244.

100. Prairie Farmer, Oct. 4, 1890, p. 637.

101. Twelfth Census, V, pt. 1, p. cliii; Wallaces' Farmer, Jan. 6, 1899, p. 13; Clay Robinson & Co., Report, Jan. 3, 1901, p. 3.

Chapter 9

1. Patent Office, Report, 1841-1861; USDA, Report, 1862-1900; Houck, Bureau of Animal Industry, pp. xi-xii.

2. Patent Office, Report, 1844, pp. 155-56, 387-89, 1012.

3. Ibid., 1861, p. 446.

4. USDA, Report, 1873, pp. 290-91.

5. Ibid., 1880, pp. 90-97.

6. Curtiss, Essentials of Beef Production, USDA Bull. 71, p. 23.

7. BAI, Report, 1884, pp. 245-69, 1885, pp. 362-76, 1887-1888, pp. 39-44, 410-13; Ritzman, Baby Beef, BAI Circ. 105, passim; BAI Circ. 106 contained a list of all bureau publications from 1884 through 1906.

8. Bierer, Short History of Veterinary Medicine, pp. 79-85; ISAST, IV:55-57, VI:xi-xv; Prairie Farmer, 1860-1861, 1866, passim.

9. Bierer, Short History of Veterinary Medicine, pp. 36-45, 79-85; Smithcors, American Veterinary Profession, pp. 394-95; Prairie Farmer, Mar. 22, 1879, p. 93, Nov. 24, 1883, pp. 737, 740, Jan. 5, 1884, p. 4, Oct. 25, 1884, p. 676, May 9, 1885, p. 292, Oct. 2, 1886, p. 646, June 11, 1887, p. 377; National Livestock Journal, Oct. 1885, p. 424.

10. Field and Farm, July 16, 1887, p. 4, Mar. 29, 1888, p. 4, Oct. 16, 1888, p. 4, Dec. 8, 1888, p. 8, and passim 1887, 1888; Chew, Response of Government to Agriculture, pp. 30-31.

11. USDA, Yearbook, 1956, pp. 263-64.

12. Ibid., 1894, pp. 67-80; Houck, Bureau of Animal Industry, pp. 25, 256-59; Smithcors, American Veterinary Profession, p. 471.

13. There were agricultural experiment stations operating without federal aid in states like Connecticut, Ohio, New York, and Wisconsin before the Hatch Act, but not in Illinois and Iowa. True, History of Agricultural Experimentation, USDA Misc. Publ. 251, pp. 67-164.

14. Northwestern Farmer, 1858, p. 5.

15. Prairie Farmer, Feb. 12, 1870, p. 41, speech to a farmers' convention, Madison County, Ill.

16. National Livestock Journal, pp. 220-22.

17. Prairie Farmer, Oct. 17, 1874, p. 333, July 3, 1875, p. 213; True, History of Agricultural Experimentation, pp. 80-82, 107-110; National Livestock Journal, Jan. 1885, pp. 20-21.

18. Ross, Land-Grant Idea, pp. 71, 90-101; Solberg, University of Illinois 1867-1894, pp. 344-48.

19. True and Clark, Agricultural Experiment Stations, USDA, Bull. 80, pp. 202-6, 532-34; University of Illinois, Agr. Exp. Sta. Bulletins, Circulars, and Reports, 1888-1900, especially, Report, 1901, p. 5; ibid., Bulletins 78, 83.

20. True and Clark, Agricultural Experiment Stations, pp. 214, 218-21, 537-40; Iowa Agricultural Experiment Station, Bulletin, 1888-1900.

21. Butchers' Advocate, Oct. 24, 1894, pp. 11-12.

22. Hamilton, History of Farmers' Institutes, USDA, Bull. 174, pp. 28-30; Prairie Farmer, Mar. 5, 1870, p. 65, Jan. 7, 1871, p. 4, Jan. 18, 1873, p. 20, Dec. 5, 1874, p. 389, Jan. 26, 1878, p. 28, Jan. 18, 1879, p. 20, Jan. 24, 1880, p. 28.

23. Hamilton, History of Farmers' Institutes, pp. 28-30; Prairie Farmer, Jan. 27, 1883, p. 56, Feb. 9, 1884, p. 89, Apr. 12, 1884, p. 227; IDAT, XX:426-29, 511-17.

24. Illinois Farmers' Institute, Report, 1895-1900, passim; ibid., 1896, pp. 264, 474.

25. Prairie Farmer, Feb. 11, 1871, p. 42, Jan. 13, 1872, p. 12, Jan. 20, 1872, p. 17; Iowa SASR, 1898-1900; Hamilton, History of Farmers' Institutes, pp. 36-38.

Chapter 10

1. Prairie Farmer, May 15, 1863, p. 313, June 13, 1863, p. 377, Aug. 29, 1863, p. 130, Apr. 30, 1864, p. 309, Sept. 3, 1864, p. 146, May 6, 1865, p. 352, Jan. 16, 1869, p. 17, Nov. 13, 1869, p. 369; USDA, Report, 1866, pp. 317-18; Throne, Southern Iowa Agriculture, Iowa J. Hist. 50:212.

2. Prairie Farmer, Oct. 17, 1861, p. 263, Dec. 12, 1861, p. 386, Nov. 1, 1862, p. 275, Oct. 10, 1863, p. 229, Apr. 15, 1865, p. 280, Nov. 17, 1866, p. 314, Jan. 15, 1869, p. 17, May 15, 1869, p. 153; USDA, Report, 1862, pp. 332-33; 1866, pp. 317-18.

3. Iowa Homestead, July 30, 1869, p. 4; Prairie Farmer, Nov. 29, 1860, p. 341, May 15, 1869, p. 153, May 29, 1869, p. 169, June 5, 1869, p. 177; Iowa SASR, 1868, pp. 216-22.

4. Iowa SASR, 1868, pp. 216-22; Prairie Farmer, Jan. 4, 1862, p. 2; Iowa Homestead, Aug. 13, 1869, p. 2.

5. Prairie Farmer, May 19, 1866, p. 337, Feb. 14, 1864, p. 98, Jan. 28, 1865, p. 51; USDA, Report, 1866, pp. 317-18; Hopkins, Economic History, pp. 115-16.

6. Prairie Farmer, Mar. 11, 1871, p. 80, May 19, 1877, p. 160.

7. Gates, Cattle Kings, Miss. Valley Hist. Rev. 35: 389-405; Hopkins, Economic History, pp. 87-95, 103, 128;

Bogue, Pioneer Farmers, <u>Iowa J. Hist.</u> 55:24; <u>Prairie Farmer</u>, Apr. 20, 1878, p. 128, Apr. 27, 1878, p. 136; Hopkins, Passing of the Herds, <u>Palimpsest</u> 11:204-5.

8. <u>Prairie Farmer</u>, Jan. 15, 1869, p. 17; Throne, Southern Iowa Agriculture, p. 212; Hopkins, <u>Economic History</u>, pp. 103, 128.

9. <u>Prairie Farmer</u>, Apr. 30, 1864, p. 309, Apr. 22, 1865, p. 304, May 5, 1865, p. 352, Apr. 29, 1871, p. 136, Sept. 6, 1873, p. 288, Sept. 13, 1873, p. 296; Bogue, Swamp Land Act, <u>Agr. Hist.</u> 25:176-77.

10. <u>Prairie Farmer</u>, Aug. 3, 1872, p. 245, June 30, 1877, p. 205; MacDonald, <u>Food from the Far West</u>, pp. 140-42; ISAST, XVI:95.

11. <u>Prairie Farmer</u>, Jan. 7, 1871, p. 8, Feb. 7, 1874, p. 48, Jan. 8, 1876, p. 16.

12. Illinois Industrial University, Fifth Annual Circular, pp. 220-22; <u>Prairie Farmer</u>, Mar. 23, 1872, p. 89.

13. <u>Prairie Farmer</u>, Aug. 10, 1872, p. 256, Sept. 14, 1872, p. 296, Jan. 11, 1873, p. 13, Feb. 8, 1873, p. 48, Apr. 5, 1873, p. 112, Sept. 1, 1877, p. 273; Chicago Board of Trade, Report, 1901, p. 12.

14. <u>Prairie Farmer</u>, Jan. 4, 1873, p. 8, Mar. 1, 1873, p. 72.

15. MacDonald, <u>Food from the Far West</u>, pp. 120-26, 137-39; <u>Prairie Farmer</u>, Feb. 27, 1876, p. 72.

16. MacDonald, <u>Food from the Far West</u>, pp. 123, 137-39; <u>Western Farm Journal</u>, May 3, 1878, pp. 274-75; <u>Prairie Farmer</u>, Feb. 25, 1871, p. 64, Feb. 1, 1873, p. 40, Feb. 8, 1873, p. 48, Mar. 24, 1873, p. 104, June 14, 1873, p. 189, Mar. 14, 1874, p. 88, Mar. 28, 1874, p. 104, Dec. 5, 1874, p. 392, Jan. 24, 1875, p. 24, Mar. 13, 1875, p. 88, Mar. 20, 1875, p. 96, Mar. 27, 1875, p. 104, Apr. 3, 1875, p. 112, June 5, 1875, p. 184, Nov. 10, 1877, p. 357, Dec. 1, 1877, p. 381, Mar. 27, 1878, p. 93.

17. Prime, <u>Model Farms</u>, pp. 470-72; see also Hopkins, <u>Economic History</u>, pp. 94, 107-8; <u>Field and Farm</u>, June 18, 1887, p. 8; <u>Iowa Register</u>, in <u>Prairie Farmer</u>, Oct. 4, 1884, p. 628.

18. Prime, <u>Model Farms</u>, pp. 17, 27-28, 36, 49-52, 63-64, 82-83, 85-86, 89-90, 92-93, 123, 262-66, 269-71, 282-84.

19. Ibid., 427-30, 443-45, 453-54, 501, 503-4, 519, 484-85.

20. <u>Prairie Farmer</u>, Aug. 14, 1880, p. 261; IDAT, XXI: 185-87.

21. Prairie Farmer, Nov. 10, 1883, p. 708.

22. IDAT, XXI:185-86; shorts were a by-product of milling wheat and contained germ, fine bran, and some flour.

23. Prairie Farmer, Aug. 15, 1883, p. 516.

24. Ibid., Nov. 14, 1885, p. 744; National Livestock Journal, Dec. 1885, p. 501.

25. Iowa Homestead, Feb. 13, 1891, p. 149, Feb. 6, 1891, pp. 122, 124, Mar. 13, 1891, p. 241, July 24, 1891, p. 689, July 31, 1891, p. 706, Dec. 11, 1891, p. 1151, Dec. 18, 1891, p. 1174; Prairie Farmer, Aug. 2, 1890, p. 484; BAI, Special Report, H. Misc. Doc. 106, pp. 456-61; Hopkins, Economic History, pp. 106-18; Wallaces' Farmer, Jan. 13, 1899, pp. 24-25, Jan. 20, 1899, p. 42, Feb. 10, 1899, p. 106, July 21, 1899, p. 595.

26. Henry, Feeds and Feeding, pp. 382-93.

27. Ibid., pp. 130-31, 152, 196, 204, 207, 225, 257-58, 367-69, 381-84; University of Illinois, Agr. Exp. Sta. Bull. 36, pp. 20, 421-31.

28. Henry, Feeds and Feeding, pp. 357-58, 367, 393.

29. Prairie Farmer, Jan. 4, 1896, p. 7, Jan. 8, 1898, p. 3, Feb. 19, 1898, p. 1, Mar. 19, 1898, p. 3, Jan. 13, 1900, p. 1; Mumford, Beef Production, pp. 57-66; Hopkins, Economic History, pp. 129-31.

30. Clay Robinson and Co., Live Stock Report, p. 3.

Chapter 11

1. U.S. Census of Agriculture, 1959, General Report, II:975, 985.

BIBLIOGRAPHY

Manuscripts

Flagg Papers, Illinois Historical Survey of the University
of Illinois, Urbana.
Hopkins Papers, Illinois State Historical Society, Spring-
field.
Little Papers, Illinois Historical Survey of the University
of Illinois, Urbana.
Scully Papers, Illinois Historical Survey of the University
of Illinois, Urbana.
Williams-Woodbury Papers, Illinois Historical Survey of the
University of Illinois, Urbana.

Federal and State Documents

Aldrich, Nelson W. Wholesale Prices, Wages, and Transporta-
tion. (52nd Cong. 2nd sess., S. Rept. 1394.) Washing-
ton, 1893.
Census of Iowa for 1880 . . . with Other Historical and Sta-
tistical Data. Des Moines, 1883.
Curtiss, Charles F. Some Essentials of Beef Production. (USDA
Farmer's Bulletin 71.) Washington, 1898.
Hamilton, John. History of Farmers' Institutes in the United
States. (USDA, Office of the Experiment Stations Bulle-
tin 174.) Washington, 1906.
Illinois Board of Live Stock Commissioners. Annual Report,
1886-1900. Springfield, 1887-1901.
Illinois Department of Agriculture. Report on the 8th Fat
Stock Show, 1885. Springfield, 1886.
Illinois Department of Agriculture. Transactions, 1871-1900.
Springfield, 1872-1901.
Illinois Farmer's Institute. Annual Report, 1895-1900.
Springfield, 1896-1901.

181

Illinois Revised Statutes, 1874.

Illinois State Agricultural Society. Transactions, 1853-1870. Springfield, 1854-1871.

Illinois State Board of Agriculture. Report of the American Fat Stock and Dairy Show, 1885. Springfield, 1886.

—————. Statistical Report for December, 1883-1900. Springfield, 1883-1900.

In Re Barber. 39 Federal Reporter 641.

Iowa State Agricultural Society. Reports, 1853-1899. Des Moines, 1854-1900.

Minnesota v. Barber. 136 U.S. Reports 313.

Nimmo, Joseph, Jr. Report on the Internal Commerce of the United States. (48th Cong. 2nd sess., H. Exec. Doc. 7, pt. 3.) Washington, 1885.

Ritzman, Ernest G. Baby Beef. (USDA, Bureau of Animal Industry Circular 105.) Washington, 1899.

Royce, Charles C., comp. Indian Land Cessions in the United States. (56th Cong. 1st sess., H. Doc. 736, pt. 2.) Washington, 1899.

Swift v. Sutphin. 39 Federal Reporter 630.

Thompson, James Westfall. A History of Livestock Raising in the United States, 1607-1860. (USDA Agricultural History Series 5.) Washington, 1942.

True, Alfred C. A History of Agricultural Experimentation and Research in the United States, 1607-1925. (USDA Miscellaneous Publication 251.) Washington, 1937.

True, Alfred C., and Clark, V. A. The Agricultural Experiment Stations in the United States. (USDA, Office of the Experiment Stations Bulletin 80.) Washington, 1900.

U.S. Census Office. Seventh Census of the United States. Compendium. Washington, 1854.

—————. Eighth Census of the United States. II, Agriculture. Washington, 1864.

—————. Ninth Census of the United States. III, Industry and Wealth. Washington, 1872.

—————. Tenth Census of the United States. III, Agriculture. Washington, 1883.

—————. Eleventh Census of the United States. VI, Agriculture. Washington, 1895.

—————. Twelfth Census of the United States. I, Population; V, Agriculture; VI, Agriculture; Abstract; Statistical Atlas. Washington, 1901-1903.

—————. Thirteenth Census of the United States. V, Agriculture. Washington, 1913.

U.S. Census Office. Seventeenth Census of the United States.
I, Population. Washington, 1952.
————. Historical Statistics of the United States, Colonial Times to 1957. Washington, 1960.
U.S. Commissioner of Agriculture. Reports, 1862-1888. Wash-
ington, 1863-1889.
U.S. Commissioner of Corporations. Report on the Beef Indus-
try. (58th Cong. 3rd sess., H. Doc. 382.) Washington,
1905.
U.S. Commissioner of the General Land Office. Report, 1838,
1843. Washington, 1838, 1843.
U.S. Commissioner of the Patent Office. Report, II, Agricul-
ture. Washington, 1841-1861.
U.S. Congress. House Executive Document 267. (48th Cong. 2nd
sess.) Washington, 1885.
————. House Document 855. (64th Cong. 1st sess.) Wash-
ington, 1916.
————. Senate. Report of the Select Committee on the
Transportation and Sale of Meat Products. (51st Cong.
1st sess., S. Rept. 829.) Washington, 1890.
U.S. Department of Agriculture, Bureau of Animal Industry.
Annual Report, 1884-1900. Washington, 1885-1901.
————. Special Report on Diseases of Cattle and on Cattle
Feeding. (52nd Cong. 2nd sess., H. Misc. Doc. 106.)
Washington, 1893.
USDA. Yearbook of Agriculture, 1940, Washington, 1940.
U.S. Federal Trade Commission. Report of the Federal Trade
Commission on Private Car Lines. Washington, 1920.
————. Report on the Meat-Packing Industry, 6 pts. and
summary. Washington, 1919-1920.
U.S. Interstate Commerce Commission. Thirteenth Annual Re-
port on the Statistics of Railways in the United States.
Washington, 1901.
U.S. Secretary of Agriculture. Report, 1889-1900. Washing-
ton, 1889-1900.
U.S. Statutes at Large, II.

Newspapers and Agricultural Periodicals

Butchers' Advocate
Breeders' Gazette
Cattleman
Drovers' Journal

Field and Farm
Iowa Homestead
National Livestock Journal
Northwestern Farmer
Prairie Farmer
Wallaces' Farmer
Western Agriculturalist
Western Farm Journal
Western Rural
"Cole Notes" in the Illinois Historical Survey of the University of Illinois. Thirty-one boxes of newspaper excerpts from 1820 to 1919. Indexed topically.

Primary Works

Allerton, Samuel W. On Systematic Farming. Chicago, 1907.

Armour, J. Ogden. The Packers, the Private Car Lines, and the People. Philadelphia, 1906.

Ayer, N. W., and Son's. American Newspaper Annual, 1880, 1891, 1900. Philadelphia, 1881, 1892, 1901.

Burlend, Rebecca; and Burlend, Edward. A True Picture of Emigration, ed. Milo Quaife. New York, 1968.

Bushnell, Horace. Work and Play: Or Literary Varieties. New York, 1871.

Bushnell, J. P. Iowa Resources and Industries: Her Agricultural, Horticultural, Stock-Raising, Dairying, Commercial, Manufacturing and Mining Interests, etc. Des Moines, 1885.

Chicago Board of Trade. Annual Report, 1858-1901. Chicago, 1859-1902.

Clay Robinson and Company. Live Stock Report. Jan. 3, 1901.

Democratic Press. The Railroads, History and Commerce of Chicago, 2nd ed. Chicago, 1854.

Drovers' Journal. Year Book of Figures of the Live Stock Trade for the Year 1922, Chicago, 1923.

Flint, Timothy. A Condensed Geography and History of the Western States, or the Mississippi Valley, 2 vols. Cincinnati, 1828.

Griffiths' Fifth Annual Review of the Live Stock Trade of Chicago for 1869. Chicago, 1870.

Hildreth, Azro B. F. The Life and Times of Azro B. F. Hildreth, 4 pts. Des Moines, 1891.

Henry, W. A. Feeds and Feeding: A Hand-Book for the Student and Stockman. Madison, 1898.

Illinois Industrial University. Fifth Annual Circular, 1871-
72. Urbana, 1872.

Iowa State College of Agriculture and Mechanic Arts, Agricultural Experiment Station. Bulletin, 1888-1900. Ames,
1888-1900.

Lea, Albert M. Notes on Wisconsin Territory, with a Map.
Philadelphia, 1836.

McCoy, Joseph G. Historic Sketches of the Cattle Trade of
the West and Southwest, ed. Ralph P. Bieber. Glendale,
1940.

MacDonald, James. Food from the Far West or, American Agriculture with Special Reference to the Beef Production
and Importation of Lean Meat From America to Great
Britain. London, 1878.

Mumford, Herbert. Beef Production. Urbana, 1907.

National Live Stock Association. Proceedings of the Fifth Annual Convention. Chicago, 1901.

Oliver, William. Eight Months in Illinois: With Information
to Immigrants. Chicago, 1924.

Peoria Board of Trade. Annual Report, 1880-1884.

A Pioneer. Northern Iowa, by a Pioneer: Containing Information for Immigrants. Dubuque, 1858.

[Ponting, Tom C.] Life of Tom Candy Ponting: An Autobiography, Introduction and notes by Herbert O. Bayer. Evan-
ston, Illinois, 1952.

Prime, Samuel T. K., ed. The Model Farms and Their Methods.
Chicago, 1881.

Schoolcraft, Henry R. Travels in the Central Portions of the
Mississippi Valley. New York, 1825.

Shirreff, Patrick. A Tour through North America; Together
with a Comprehensive View of the Canadas and United
States, as Adapted for Agricultural Emigration. Edin-
burgh, 1835.

Union Stock Yards and Transit Company. Annual Live Stock Report, 1868-1901. Chicago, 1869-1902.

——————. Review of the First International Live Stock Expo-
sition, 1900. Chicago, 1901.

University of Illinois, Agricultural Experiment Station. Annual Report. Urbana, 1888-1910.

——————. Bulletin 9, 36, 46. Urbana, 1890, 1894, 1897.

——————. Circular. Urbana, 1897-1902.

Wistar, Isaac Jones. Autobiography of Isaac Jones Wistar
1827-1905. New York, 1937.

Secondary Works

Anderson, Oscar E., Jr. Refrigeration in America: A History
    of a New Technology and Its Impact. Cincinnati, 1953.

Andreas, A. T. History of Chicago, 3 vols. Chicago, 1884.

Angle, Paul M., comp. and ed. Prairie State: Impressions of
    Illinois, 1673-1967, by Travelers and Other Observers.
    Chicago, 1968.

Bardolph, Richard C. Agricultural Literature and the Early
    Illinois Farmer. (University of Illinois Studies in the
    Social Sciences, 29.) Urbana, 1948.

Bateman, Newton; and Selby, Paul, ed. Historical Encyclopedia
    of Illinois and History of Morgan County. Chicago, 1906.

Bierer, Bert W. A Short History of Veterinary Medicine in
    America. East Lansing, 1955.

Bogue, Allan G. From Prairie to Corn Belt: Farming on the
    Illinois and Iowa Prairies in the Nineteenth Century.
    Chicago, 1963.

Cavanagh, Helen M. Funk of Funk's Grove: Farmer, Legislator
    and Cattle King of the Old Northwest 1797-1865. Bloom-
    ington, Illinois, 1952.

Chew, Arthur P. The Response of Government to Agriculture.
    Washington, 1937.

Clark, John G. The Grain Trade in the Old Northwest. Urbana,
    1966.

Clemen, Rudolf A. The American Livestock and Meat Industry.
    New York, 1923.

Cole, Arthur C. The Era of the Civil War 1848-1870. (Centen-
    nial History of Illinois, vol. 3.) Springfield, 1919.

Conway, H. M. Cattle Handbook for the Grower and Feeder.
    Chicago, 1935.

Dale, Edward E. The Range Cattle Industry: Ranching on the
    Great Plains from 1865 to 1925. Norman, 1960.

Danhof, Clarence H. Change in Agriculture: The Northern
    United States, 1820-1870. Cambridge, 1969.

Dykstra, Robert R. The Cattle Towns. New York, 1968.

English, William Hayden. Conquest of the Country Northwest
    of the River Ohio 1778-1783 and the Life of Gen. George
    Rogers Clark, vol. 1. Indianapolis, 1896.

Fenneman, Nevin M. Physiography of the Eastern United States.
    New York, 1938.

Garland, John H., ed. The North American Midwest: A Regional
    Geography. New York, 1955.

Gates, Paul W. The Farmer's Age: Agriculture, 1815-1860. New
    York, 1960.

Gates, Paul W. _The Illinois Central Railroad and Its Colonization Work._ (Harvard Economic Studies, 42.) Cambridge, 1934.

Gressley, Gene M. _Bankers and Cattlemen._ New York, 1966.

Grodinsky, Julius. _The Iowa Pool: A Study in Railroad Competition, 1870-1884._ Chicago, 1950.

Hayter, Earl W. _The Troubled Farmer 1850-1900._ De Kalb, Illinois, 1968.

Henlein, Paul C. _Cattle Kingdom in the Ohio Valley 1783-1860._ Lexington, 1959.

History of Morgan County, Illinois: Its Past and Present, Containing a History of the County; Its Cities, Towns, etc. Chicago, 1878.

Hopkins, John A., Jr. _Economic History of the Production of Beef Cattle in Iowa._ Iowa City, 1928.

————. _A Statistical Study of the Prices and Production of Beef Cattle._ (Iowa State College of Agriculture and Mechanic Arts, Agricultural Experiment Station Research Bulletin 101.) Ames, 1926.

Houck, U. G. _The Bureau of Animal Industry of the United States Department of Agriculture: Its Establishment, Achievements and Current Activities._ Washington, 1924.

Kay, George F., et al. The Pleistocene Geology of Iowa. (Iowa Geological Survey, Special Report.) N.p., n.d., c/1944.

Lampard, Eric. _The Rise of the Dairy Industry in Wisconsin: A Study in Agricultural Change 1820-1920._ Madison, 1963.

Leech, Harper; and Carroll, John Charles. _Armour and His Times._ New York, 1938.

Lokken, Roscoe L. _Iowa Public Land Disposal._ Iowa City, 1942.

Lord, Russell. _The Wallaces of Iowa._ Boston, 1944.

Morris, Ira Nelson. _Heritage from My Father: An Autobiography._ New York, 1947.

Odell, R. T., et al. Soils of the North Central Region of the United States: Their Characteristics, Classification, Distribution, and Related Management Problems. (University of Wisconsin, Agricultural Experiment Station Bulletin 601.) Madison, 1960.

Ogilvie, William E. _Pioneer Agricultural Journalists._ Chicago, 1927.

Pierce, Bessie Louise. _A History of Chicago_, 3 vols. Chicago, 1937-1957.

Pooley, William Vipond. The Settlement of Illinois from 1830 to 1850. (University of Wisconsin Bulletin 220.) Madison, 1908.

Power, Richard L. Planting Corn Belt Culture: The Impress of the Upland Southerner and Yankee in the Old Northwest. Indianapolis, 1953.

Putnam, James W. The Illinois and Michigan Canal: A Study in Economic History. Chicago, 1918.

Rogin, Leo. The Introduction of Farm Machinery in Its Relation to the Productivity of Labor in the Agriculture of the United States during the Nineteenth Century. Berkeley, 1931.

Ross, Earle D. Iowa Agriculture: An Historical Survey. Iowa City, 1951.

—————. The Land-Grant Idea at Iowa State College. Ames, 1958.

Ross, R. C.; and Case, H. C. M. Types of Farming in Illinois: An Analysis of Difference by Areas. (University of Illinois, Agricultural Experiment Station Bulletin 601.) Urbana, 1956.

Shannon, Fred A. The Farmer's Last Frontier: Agriculture, 1860-1897. New York, 1945.

Smithcors, J. F. The American Veterinary Profession: Its Background and Development. Ames, 1963.

Solberg, Winton U. The University of Illinois 1867-1894: An Intellectual and Cultural History. Urbana, 1968.

Swift, Louis F., with Arthur Van Vlissingen, Jr. The Yankee of the Yards: The Biography of Gustavus Franklin Swift. Chicago, 1927.

Taylor, George Rodgers; and Neu, Irene. The American Railroad Network, 1861-1890. Cambridge, 1956.

Van Stuyvenberg, J. H., ed. Margarine: An Economic, Social and Scientific History 1869-1969. Toronto, 1969.

Wallace, Henry A.; and Bressman, Earl N. Corn and Corn Growing, 5th ed., rev. New York, 1949.

Articles

Agnew, Dwight L. The Rock Island Railroad in Iowa, Iowa Journal of History 52(July 1954):203-22.

Baldwin, W. W., ed. Driving Cattle from Texas to Iowa, 1866, Annals of Iowa 14(April 1924):243-62.

Beinhauer, Myrtle. The County, District, and State Agricultural Societies of Iowa, Annals of Iowa 20(July 1935): 50-69.

—————. Some Early Iowa Agricultural Organizations, Annals of Iowa 34(October 1958):422-33.

188

Bogue, Allan G. Pioneer Farmers and Innovation, _Iowa Journal of History_ 55(January 1958):1-36.

Bogue, Margaret B. The Swamp Land Act and Wet Land Utilization in Illinois, 1850-1890, _Agricultural History_ 25(October 1951):169-80.

Buck, Solon J. Pioneer Letters of Gershom Flagg, _Transactions of the Illinois State Historical Society, 1910_, pp. 139-83.

Buckingham, Clyde E. Early Settlers of the Rock River Valley, _Journal of the Illinois State Historical Society_ 35(September 1942):236-59.

Carman, Harry J. English Views of Middle Western Agriculture, 1850-1870, _Agricultural History_ 8(January 1934):3-19.

Chandler, Alfred D., Jr. The Beginnings of "Big Business" in American Industry, _Business History Review_ 33(Spring 1959):1-31.

Dunbar, Gary S. Colonial Carolina Cowpens, _Agricultural History_ 35(July 1961):125-31.

East, Ernest E. The Distillers' and Cattle Feeders' Trust, 1887-1895, _Journal of the Illinois State Historical Society_ 45(Summer 1952):101-23.

Gates, Paul W. Cattle Kings in the Prairies, _Mississippi Valley Historical Review_ 35(December 1948):379-412.

————. Frontier Landlords and Pioneer Tenants, _Journal of the Illinois State Historical Society_ 38(June 1945):143-206.

Goodwin, Cardinal. The American Occupation of Iowa, 1833 to 1860, _The Iowa Journal of History and Politics_ 17(January 1919):83-102.

Harper, C. A. The Railroad and the Prairie, _Transactions of the Illinois State Historical Society, 1923_, pp. 102-10.

Harris, Mary V., ed. The Autobiography of Benjamin Franklin Harris, _Transactions of the Illinois State Historical Society, 1923_, pp. 72-101.

Hopkins, John A., Jr. The Passing of the Herds, _Palimpsest_ 11(July 1930):282-91.

Johnson, Curtis L. E. S. Ellsworth: Iowa Land Baron, _Annals of Iowa_ 35(July 1959):1-35.

Johnston, William J. Sketches of the History of Stephenson County, Illinois, _Transactions of the Illinois State Historical Society, 1923_, pp. 217-320.

Kerrick, L. H. Life and Character of Honorable Isaac Funk, _Transactions of the Illinois State Historical Society, 1902_, pp. 159-70.

Kujovich, Mary Y. The Refrigeration Car and the Growth of

the American Dressed Beef Industry, <u>Business History Review</u> 44(Winter 1970):460-82.

Le Duc, Thomas. Public Policy, Private Investment, and Land Use in American Agriculture, 1825-1875, <u>Agricultural History</u> 37(January 1963):3-9.

Lee, Judson F. Transportation—A Factor in the Development of Northern Illinois Previous to 1860, <u>Journal of the Illinois State Historical Society</u> 10(April 1917):17-85.

McClelland, Clarence P. Jacob Strawn and John T. Alexander: Central Illinois Stockmen, <u>Journal of the Illinois State Historical Society</u> 34(June 1941):177-208.

Murray, Janette S. Shipping the Fat Cattle, <u>Palimpsest</u> 28 (March 1947):85-94.

Newcombe, Alfred W. Alson J. Streeter—An Agrarian Liberal, <u>Journal of the Illinois State Historical Society</u> 38(December 1945):414-45; 39(March 1946):68-95.

Perrin, Richard. The North American Beef and Cattle Trade with Great Britain 1870-1914, <u>The Economic History Review</u>, 2nd Ser., 24(August 1971):430-44.

Porter, John L. History of Harmon Township, Lee County, Illinois, <u>Journal of the Illinois State Historical Society</u> 10(January 1918):593-638.

Rammelkamp, C. H., ed. The Memoirs of John Henry: A Pioneer of Morgan County, <u>Journal of the Illinois State Historical Society</u> 18(April 1925):39-75.

Sanders, Alvin H. James N. Brown as a Breeder and Importer of Live Stock, <u>Transactions of the Illinois State Historical Society, 1912</u>, pp. 170-75.

Throne, Mildred. "Book Farming" in Iowa, 1840-1870, <u>Iowa Journal of History</u> 49(April 1951):117-42.

————. Southern Iowa Agriculture, 1865-1870, <u>Iowa Journal of History</u> 50(July 1952):212.

————. The Burlington and Missouri Railroad in Iowa, <u>Palimpsest</u> 33(January 1952):1-32.

————. A Population Study of an Iowa County in 1850, <u>Iowa Journal of History</u> 57(October 1959):317-19, 321.

Wentworth, Edward N. Some Observations on Constructive Agricultural Movements, <u>Agricultural History</u> 27(April 1953): 41-48.

Yungmeyer, D. W. An Excursion into the Early History of the Chicago and Alton Railroad, <u>Journal of the Illinois State Historical Society</u> 38(March 1945):7-37.

Zimmerman, William David. Live Cattle Export Trade between United States and Great Britain, 1868-1885, <u>Agricultural History</u> 36(January 1962):46-52.

Theses and Dissertations

Anderson, Russell H. "Agriculture in Illinois during the Civil War Period, 1850-1870." Ph.D. dissertation, University of Illinois, 1929.

Brinkman, Leonard. "Historical Geography of Improved Cattle in the United States to 1870." Ph.D. dissertation, University of Wisconsin, 1964.

Harding, Jennie L. "Economic and Social Conditions in Iowa to 1880 as Reflected in the Observations of Travelers." M.S. thesis, Iowa State College of Agriculture and Mechanic Arts, 1942.

Hawk, Duane C. "Iowa Farming Types: 1850-1880." M.A. thesis, State University of Iowa, 1957.

Massey, Robert E. "British Travelers' Impressions of Agriculture in Illinois, 1800-1860." M.A. thesis, University of Wisconsin, 1963.

Nelson, Peter. "A History of Agriculture in Illinois with Special Reference to Types of Farming." Ph.D. dissertation, University of Illinois, 1930.

Newlove, George H. "Economic History of Illinois Agriculture." Ph.D. dissertation, University of Illinois, 1917.

Smith, Marjorie C. "Historical Geography of Champaign County, Illinois," Ph.D. dissertation, University of Illinois, 1957.

SUBJECT INDEX

[County Index follows Subject Index]

193

196

State Board of Agriculture, 92
State College of Agriculture and Mechanic Arts, 111-12, 114
State Fair, 84
Iowa Homestead, 57, 69, 71, 101, 103, 109
founding and development, 86, 90, 93

J

John, Alexander, 81
Jungle, The, 110

K

Kansas, 55, 57, 62
Kansas City, 41, 43, 46, 53-54, 57, 126
as market, 56-57, 73
Kellogg, O. W., 30
Kentuckians, 79
Kentucky, cattle feeding, 18, 79
Kerrick, L. H., 71
Kilbourne, F. L., 61

L

Land
drainage, 77
in farms, 15
surveys, 8
values, 20, 77, 83
Lawrence, E. S., 111
Lea, Albert Miller, 7
Le Duc, William G., 109
Libby, McNeil, and Libby, 49
Liverpool, England, 51

Loan companies, cattle, 65
Logan, John A., Jr., 97

M

McCoy, Joseph, 22, 59, 60, 62, 83
Machinery, farm, 14-15
Mahanah, Dennis, 82
Market, cattle. See also Baltimore, Boston, Chicago, Kansas City, New Orleans, New York City, Omaha, Philadelphia
conditions, 30-33
grades, 68
numbers, 43
reports, 31, 67, 89-90
Mason, Charles, 21
Meat. See also Dressed beef
canned, 49
distribution, 47-48
inspection, 106, 110
packers. See also Armour and Company; Morris, Nelson; Swift and Company
as cattle feeders, 52
decentralization, 53-54
influence on rail rates, 48
price fixing, 52-53, 71-72
processing. See Barreled beef, Dressed beef, Pork packing
Michaels, P. E., 77
Miller, Mark, 86
Miller, T. L., 96
Mills, Oliver, 81, 99-100
Minnesota, 55, 63
Missouri, 57-59, 62
Montana, 57
Morrill Act (1862), 110-11
Morris, Nelson, 47, 50, 52-53, 109

197

Morrow, George, 108
Mumford, Herbert W., 113
Myrick, Willard, 41

N

National Cattle Growers'
    Association, 48
National Livestock Bank, 43
National Livestock Journal,
    90, 105, 111, 119
Nebraska, 55, 57, 62
New Genesee Farmer, 25
New Orleans, 11, 19, 30, 79
New York City, 34, 53, 66,
    79
  beef consumption, 40
  markets, 18, 21, 31
New York Tribune, 20, 31
Nofsinger and Company, 46

O

Oakwood, J. H., 84
Oats, 15-16
Ohio
  cattle feeding, 18-19, 33,
    55
  Dairymen's Protective As-
    sociation, 98
  Stock Raising Company, 101
Oklahoma City, 53
Oleomargarine, 97-98
Omaha, 43, 46, 53-54, 57,
    126
Ottumwa, Iowa, 39

P

Patent Office
  advice on feeding, 25

agricultural reports, 106-
    7
Periodicals, agricultural,
    52, 85-91, 99, 101. See
    also Breeders' Gazette,
    Iowa Homestead, Prairie
    Farmer
  abuse of Bureau of Animal
    Industry, 109
Philadelphia, 18, 27, 40, 46
Ponting, Tom C., 22, 24, 119
Pool. See Price fixing
Population, 7-8, 11, 15-17,
    122
Pork packing, 36, 41
Prairie, 3-5, 7-8, 10-11, 19,
    21, 116
Prairie Farmer
  cattle raising, 23, 32, 56-
    57, 65, 94-96, 102-4,
    111
  commission firms, 67-68
  cooked food, 25
  correspondents, 20, 40
  dressed beef, 38, 40, 46-
    48, 51
  founding and development,
    85-90
  live trade, 50
  market reports, 72, 117
  shelter, 28-29
Price fixing, 52-53
Prices
  agricultural, 125
  beef, investigation, 71-72
  cattle, 30-33, 42-43, 55,
    62-64, 70-72, 115
  corn, 20
  meat, 52
Prince, Frederick, 42
Profits, cattle, 27-28, 32-
    33, 78, 86-87, 126
Purebreds, 78, 81-83, 91-94,
    100-102

Swine, 3, 10-12, 14-16, 18, 26, 28, 41, 115-16, 126

C O U N T Y   I N D E X